Handheld Radio Field Guide

Front Panel Programming (FPP) Instructions
for Handheld Ham Radios (HTs)

Second Edition

Handheld Radio Field Guide

Front Panel Programming (FPP) Instructions
for Handheld Ham Radios (HTs)
Second Edition

By Andrew Cornwall

Listening Bird Press

Handheld Radio Field Guide
Front Panel Programming (FPP) Instructions
    for Handheld Ham Radios (HTs)
    Second Edition

Text and images copyright © 2018, 2020 by Andrew Cornwall.

Published by Listening Bird Press.

ISBN 978-0-9996609-1-1

2 3 4 5 6 7 8 9 0 – 22 21 20

Contents

Introduction

There are many different handheld ham radios available today. Your choice will be based on many factors, including features, price, durability and ease of use. Because there is no standard for programming a handheld radio, the steps you'll need to take in order to enter frequencies into your radio will depend on the radio you've bought.

Today, many people purchase programming cables and software to get their radios programmed. They clone a friend's radio or get the frequencies programmed "at the store." But if you don't know how to enter frequencies directly from the front panel, you can run into problems in the field. You'll lose frequency agility—an advantage unique to the ham radio service, and one of the reasons that hams are valuable in emergency situations.

This guide has a simple purpose: to describe how to program handheld radios in the field. You will be able to program a frequency, CTCSS tone and offset into the radio, then transfer that programming to a memory so another radio operator can use it. If ham radio is deployed at an event or in an emergency, this may be enough to get you (or someone else) going when an existing repeater frequency has to change because of interference or range issues.

To be included in this guide, a radio must have programmable memories for both frequency and CTCSS tones on transmit. Only radios which can be programmed in the field on their own—without using a computer or programming cable—are included.

Each entry has a standard format. First, there is a picture of the radio layout with interesting buttons marked. Then a few specifications for the radio are given (including transmit and receive frequency ranges). Next is a list of the standard tasks you might do with the radio. This includes programming a frequency to a memory in the field, locking/unlocking the radio, checking a repeater input frequency, changing power levels in the field, adjusting volume and adjusting squelch.

Many radios have modes that are easy to get into but hard to get out of. In the Weird Modes section, you may find assistance with this issue. At the end, you'll find other useful information about the radio, including how to reset it when nothing else works.

Entries are sorted lexicographically, not numerically or by date of manufacture. For instance, this means that the Yaesu FT-252 comes before the Yaesu FT-60.

The instructions provided for programming the radio may not be the most efficient. Instead, the instructions are as bulletproof as possible to users unfamiliar with the radio. For instance, with most of the Chinese radios, this guide has the user enter Menu mode, set the value and then exit Menu mode (which stores the setting). It's possible to chain multiple settings without leaving Menu mode on many of these radios. Feel free to take any shortcuts that your radio supports.

With many of the more modern radios, firmware updates are frequent. Menus might have moved around or changed number since the guide entry was written. If you don't see what you expect, try looking near that area for the appropriate function.

If your handheld isn't in the guide, look at other radios made by the same manufacturer at around the same time. Often manufacturers made multiple variants (e.g., 2 m and 70 cm) that were programmed identically.

Manufacturers usually sell similar radios for different IARU regions. Often there is an "A" version for the Americas (Region 2) and an "E" version for Europe (Region 1). Outside of the Americas, most of the 2 m radios will transmit from 144–146 MHz; most of the 70 cm radios will transmit from 430–440 MHz. Transmission on 1.25 m (220 MHz) is generally omitted on radios outside Region 2. Region 1 radios may have a button to transmit a 1750 Hz tone used to bring up European repeaters; sometimes this replaces the Monitor button. Radios in Region 2 have reception of mobile telephone frequencies blocked because of U.S. legislation.

Speaking of legalities, it's hard to talk about radios without using their names. All trademarks used in this book are properties of their respective owners.

⚠ **WARNING:** The information in this guide is intended to help you operate your radios in the field. It is provided "as-is" without any warranty. It's possible something in this guide may damage your radio. When in doubt, refer to the original user manual.

A Plea for Sanity

This book shouldn't exist. Radios today have enough horsepower and display capability to implement a standard method of programming. It seems as if there are two reasons this hasn't been done: the lack of a standard and lack of desire among radio manufacturers.

This book can't change the radio manufacturers, but it can suggest the following standard for radio programming.

All radios should have a button labeled **PROGRAM**. Pressing it should start a sequence of prompts to the user for all the parameters necessary for programming a memory:

1. the receive frequency

2. the transmit frequency

3. the transmit tone type (CTCSS, DCS or none)

4. the transmit tone value (skipped if there is no transmit tone type)

5. the receive tone type (CTCSS, DCS or none)

6. the receive tone value (skipped if there is no receive tone type)

7. the transmit power level

8. the memory location into which the values should be stored.

After the user enters the appropriate value, the **PROGRAM** button should move to the next step, until the memory is programmed. At that point, the radio should enter memory mode with the newly-added memory selected.

The user should be able to turn the radio off to abort programming. Radios with small displays should use scrolling text to prompt.

Today's radios are great pieces of hardware—but their firmware sometimes isn't quite as polished.

What is a VFO Anyway?

Amateur radio has a lot of terminology that may be unfamiliar to new people. Some terms come from the commercial radio industry, some from electronics, and some are unique to amateur radio.

It's not possible to explain it all—that would be a book in itself—but here are a few common words and phrases that are used in this book, with explanations for those who might not be in the know.

ARS Automatic repeater shift. The feature in some radios to set the shift to an "expected" value based on the frequency you enter in the VFO.

barrel connector Many handheld radios have a way to connect them to a power supply for charging. The most common connector used is a coaxial power connector, commonly called a barrel connector. These connectors have different inner and outer sizes.

C4FM A relatively new digital voice mode introduced in 2013 by Yaesu, proprietary and exclusive to some Yaesu radios. C4FM repeaters can be configured to receive both analog FM and digital C4FM mode, and transmit analog out.

center positive When a radio uses a coaxial power connector, it can be configured in one of two ways. The most common way is with positive voltage connected to the inner part of the connector, and negative voltage connected to the outer part of the connector. This is the center positive configuration. Some radios use a *center negative* configuration instead. Connecting a center positive connector to a center negative radio or vice versa can cause a fuse to blow or even damage the radio.

channel This is a term used by many Chinese radios for *memory*.

CTCSS tone Continuous Tone Coded Squelch System. Most repeaters will not retransmit audio unless they also receive a signal that indicates the transmission is intended for that repeater. This helps reduce unintended transmissions when two repeaters on the same frequency hear the same signal. A simple way to do this is to introduce a subaudible tone which is played continuously while a radio transmits. Each repeater will then retransmit only when the appropriate tone is present. You might also hear this called a "PL" tone, which is a Motorola trademarked name for this system.

DCS Digital coded squelch. A more complex form of squelch than CTCSS. This system sends different tones in a digital pattern to open squelch, rather than a single tone. Occasionally called "CDCSS."

DMR Digital Mobile Radio. A commercial standard for digital voice which has been widely adopted commercially in North America, and adapted for amateur use. This mode has a standard, European Telecommunications Standards Institute TS 102 361.

D-STAR A digital standard for voice designed by the Japan Amateur Radio League and adopted by Icom. Recently, Kenwood has released a radio with D-STAR capabilities.

DTMF Dual-Tone Multi-Frequency. Many radios allow these touch-tone signals to be sent while pressing a number key during transmit. These tones can be used to control some repeaters.

dual watch Radios with a single receiver can receive only one frequency at a time. Some radios simulate two receivers by switching the single VFO rapidly between two frequencies. If a signal is detected, the radio stops switching frequencies and remains on the frequency with the signal.

Fusion See *C4FM*.

IARU region The International Amateur Radio Union defines three regions: Region 1 is Europe, Africa, Middle East and Northern Asia; Region 2 is the Americas; Region 3 is Asia Pacific. These regions conform to the International Telecom Union regions, so you might see them referred to as ITU regions.

lock Many radios offer the ability to prevent users from making unintentional changes to the configuration. Most offer a keyboard lock which causes the radio to ignore most keyboard buttons (with the notable exception of PTT). Some radios also offer a PTT lock and/or a knob lock, so the radio will ignore the PTT button or knob changes.

memory Modern radios offer a way to store and return to frequencies in a memory location on the radio. Often, radios will store other parameters in the memory, so they don't need to be set every time: shift, offset, squelch tones, sometimes power level. Memories may include a user-friendly name.

memory bank A way to group memories together. This is useful when scanning, and also to keep memories which are used for the same task within easy reach of each other.

offset When a radio is set to communicate through a repeater, it will be tuned to one frequency when transmitting, and then automatically retune itself to a different frequency when receiving. The difference in these two frequencies is the offset.

Part 90 The section of Federal Communications Commission regulations for commercial radios. In the United States, it is legal to use commercial radios in amateur bands.

power The strength with which a radio amplifies its transmit signal, usually measured in watts. Higher power levels will go longer distances, but use the battery faster. A radio's power level has no influence on how well it receives.

priority channel A special form of *dual watch*. Rather than switching between two arbitrary frequencies, the radio will switch between the current frequency and a previously-defined frequency. Radios will often check the priority frequency for activity even if there is activity on the other frequency.

PTT Push To Talk. A button on the radio which causes it to go into transmit mode when pressed, and into receive mode when released.

repeater A combination transmitter/receiver which retransmits what the receiver hears. Repeaters are often used for longer-distance communications between radios.

reset A way to revert the values in a radio to the original values before the radio was set up. Many radios offer ways to reset portions of the radio (a *partial reset*) rather than the entire radio (*full reset*).

step The minimum amount by which a frequency can change on the radio. For instance, a step size of 5 kHz might allow you to tune 147.000 MHz, 147.005 MHz, 147.010 MHz, 147.015 MHz.... Sometimes you will need to change the step size to select the frequency you want; you might have to use a step size of 6.25 kHz or 12.5 kHz to tune 147.0125 MHz.

shift In repeater operation, the shift determines if the offset is added to the receive frequency or subtracted from the receive frequency to yield the transmit frequency. A positive shift indicates the transmit frequency is greater than the receive frequency; a negative shift indicates the transmit frequency is less than the receive frequency.

simultaneous receive Some radios have two receivers built into the radio. These radios offer simultaneous receive: the ability to listen to transmissions on two frequencies at the same time.

squelch The ability of the radio to mute receive audio. *Carrier squelch* will mute the audio if no signal is present. *CTCSS squelch* will mute the audio if a subaudible tone is not present in the received signal. *DCS squelch* will mute the audio unless the received signal includes the correct digital data.

VFO Variable Frequency Oscillator. The component which alters the radio's frequency. Most handheld radios have one VFO; those with simultaneous receive have two. Fixed-frequency crystal-controlled radios do not have a VFO. Software-defined radios may simulate a VFO in software. Entering a frequency directly is often called "VFO mode."

Dead Battery Blues

Most early handhelds used nickel cadmium battery packs. These batteries discharged while on the shelf after about 30 days, and could handle about 500 charge cycles.

When the nickel metal hydride chemistry came along, users appreciated the greater energy densities. The improvement in energy density offset the disadvantage of a reduced number of charge cycles: NiMH cells can take about 300 charges before no longer being able to perform.

The transition from NiCd to NiMH was fairly smooth. Manufacturers found that NiMH batteries could be trickle-charged in the same way that the previous NiCd chemistry could.

The advent of lithium ion batteries presented a problem to manufacturers. Li-ion batteries have a much higher power density than NiMH batteries and have a lower self-discharge rate. They can be charged about the same number of times as NiMH batteries: 300 charge cycles. However, li-ion batteries need more advanced charging, and can't use the same trickle-charging circuitry that the nickel-based batteries did, which creates an incompatibility.

These days, almost all new radios use lithium ion batteries and have a charger designed for that. In the transition period from NiCd and NiMH to lithium ion, some manufacturers provided two chargers—one for the nickel chemistries, and one for the lithium ion chemistry. Then they made design changes to the battery packs to ensure the batteries were charged correctly. These changes were sometimes mechanical: additional plastic tabs and slots to prevent the battery from physically being inserted in the wrong charger, or charging pads at different locations than nickel-based batteries. Sometimes the changes were electronic: the radio had a menu to select between nickel batteries or lithium batteries and relied on the user to set it appropriately. In short, similar-looking battery packs and chargers may not work with each other.

There's not much you can do in the field if someone shows up with a dead battery. In rare cases, you might be able to rig a charger, *if* the radio has a charging jack or the user brought the correct charger.

Knowing about chargers for different battery chemistries may help a clueless user on future events. Possibly more helpful is a recommendation to buy an AA or AAA battery pack for the radio that can be used in place of the rechargable battery—and which will keep a radio powered long after the manufacturer stops production of the original batteries.

The Radio Has No Buttons

Before front panel programming, handheld radios had other methods of frequency entry. Some radios had no buttons at all. Even though radios like the Icom IC-2A had buttons on the front panel, they weren't for frequency entry—they just sent DTMF tones.

A radio before front panel programming. In this case the lack of an LCD screen and alternate function labels on the buttons suggest it doesn't have memories.

Setting the frequency

Early handheld radios had a single crystal-controlled frequency. There was no way to change those frequencies except by opening up the radio and swapping crystals. Slightly later, radios held more than one crystal, and frequencies were set via a switch (usually A/B or 1/2, sometimes A/B/C or 1/2/3 for radios that accommodated three crystals). Again, if the frequency you wanted wasn't one of the crystal frequencies, you'd have to open the radio up and swap crystals. If you're faced with a radio like this, the odds of getting it to work with a modern repeater are slim unless you routinely carry crystals around.

After the advent of phase lock loop (PLL) tuning, handheld radios had direct frequency entry, typically by setting thumbwheel switches on the top of the radio:

IC-2AT set to 144.390 MHz

These radios are almost always single band. You have to know which band the radio operates in. This is often but not always embedded in the model number. Because these radios are single band, users could assume the first two digits were "14" for 2 m receivers, "22" for 220 MHz receivers, and "44" for 70 cm receivers. Users then set the remaining digits using the thumbwheels.

On some radios, the thumbwheels simply add to the base frequency (144 on 2 m, 220 on 220 MHz, and 440 on 70 cm). In these cases, you might set the thumbwheels to "039" to specify 144.390, and "324" to specify 147.240. The IC-2AT is more sophisticated: if the first thumbwheel is 0, 4 or 8 it uses 144 MHz; if the first thumbwheel is 1, 5 or 9 it uses 145 MHz; if the first thumbwheel is 2 or 6 it uses 146 MHz; if the first thumbwheel is 3 or 7 it uses 147 MHz. European variants which don't include 146–147 MHz will allow selection of 144–145 MHz only.

In addition to the thumbwheels, this radio has a +5 kHz switch. When this switch is turned off, the radio's frequency is what is set on the thumbwheel switches. When this switch is turned on, the radio tunes to 5 kHz higher than what the thumbwheels read (allowing a user to tune to 144.395 MHz, for instance).

Repeater operation

Radios which were made after repeaters became prevalent have another set of switches:

IC-2AT duplex and power settings

These switches determine whether the transmit frequency is different from the receive frequency, and if so by how much. (The output power level switch is also here for this transceiver.)

On radios that know about repeaters, the frequency you're setting using thumbwheel knobs is usually the receive frequency—but sometimes it's the transmit frequency.

For the most part, these radios do not allow CTCSS encoding. Some optional boards exist to provide tone encoding. Older boards transmit one CTCSS tone only. Newer boards can transmit different tones; they typically use DIP switches, jumpers, solder pads or a trim potentiometer to set the tone. Very recent tone boards include a digital display and buttons to set the tone.

Some tone boards have a separate potentiometer for CTCSS volume control.

A typical tone board with volume control

Solder pads set the frequency

When installed, tone boards are typically inside the body of the radio. You'll often need to open the radio to set the CTCSS tones, if you have that option.

I Don't Speak Chinese! (or Japanese, or Korean, or...)

Many radios have the ability to show their user interface in multiple languages. This allows manufacturers to use the same radio for different markets without having to flash different software for each market.

That's normally not a problem until you actually have to reset the radio. At that point, the radio may revert to a language that you don't speak. The smarter manufacturers have made the procedure for changing languages multilingual—for instance, some radios have both English and Chinese words for `Language` in their menu systems. Not everyone does that, though.

Even once you've found the "Language" section, you may hit a roadblock. Good systems show the word for "English" in English, the word for "Japanese" in Japanese, and the word for "Chinese" in Chinese. But some manufacturers take a different approach: if you're in English, the word for "Chinese" is just that: `Chinese`. That's no help to a native Chinese speaker. Similarly, if you're in Chinese, the word for "English" might be 英文寫作 (the Chinese characters that mean "English writing") rather than the word `English`.

Here are some common terms in different languages that you may encounter. If you have to reset a radio and it reverts to a non-English language, this may help you get back to something you can read.

In general, a procedure you can use is to look in the menu system for something that means "Language" and then (often one level below that in the menu system) something that means "English." Sometimes there is a menu above that—usually "System" or "Configuration." Things get easier if you have a similar radio handy which has not been reset.

If you're confronted with multiple possible options, it's best to write down which ones you have tried. This is easier to do if the menu system has numbers. If not, remember how many button pushes it took to get to that menu entry, and hope that the menus are not alphabetized!

Chinese

In a Chinese menu system, you might look for a character included in "writing" 文 or a character included in "language" 语. From there you'd look for something including the character 英 (part of "English").

Japanese

In Japanese, you should look for something that has 語 in it (part of "language") and then look for something that has 英 (part of "English") in it. This character used in "English" is similar in Japanese and Chinese.

Korean

In Korean, you want something that includes at least one of the two characters 언어 in it ("language") and then look for something that looks like 영어 ("English").

Russian

In Russian, you'll see something like язык ("language"). Under that, look for английский ("English").

Other European Languages

If for some reason your radio defaults to Spanish, German, or French, look for a menu that contains "idioma," "Sprache" or "langue" ("language"). Then look for "Inglés," "Englische" or "Anglais(e)" ("English").

Who made my radio?

You may have a radio that looks very similar to one in this book but which carries a badge indicating that it was made by a different manufacturer. There are a number of reasons that this might happen:

- The manufacturer might have been purchased by another company while the radio was in production.

- The manufacturer may have chosen different branding to make the product sell better.

- The company selling the radio may have purchased it from the manufactuer and then re-badged it as their own, sometimes with minor firmware or hardware changes. Menus numbers may differ slightly.

In addition, some models have been renamed/renumbered to avoid unpleasant connotations or just to align with different product lines in different geographies.

Regardless of the reason, this can make it harder to find your radio. To make things a little simpler, we provide this non-exhaustive cross-reference of radios which are programmed the same.

If you see	Look under
Alan CT-22	Midland 73-030
BTech DMR-6X2	Anytone AT-D868UV
BTech UV-5X3	Baofeng UV-5R
Cignus UV-82	Baofeng UV-5R
EDCgear TH-UV7R	Baofeng UV-5R
Maha RL-115	Midland 73-030
Midland CT-22	Midland 73-030
Misuta UV-82	Baofeng UV-82
Pofung	Baofeng
Racal 25	Thales 25
Realistic	Radio Shack
Retevis RT-5	Baofeng UV-5R
Rexon RL-115	Midland 73-030
Rugged Radios RH-5R	Baofeng UV-5R
Tenway UV-5R	Baofeng UV-5R
Tidradio TD-F9GP	Baofeng UV-5R

Radio Layout

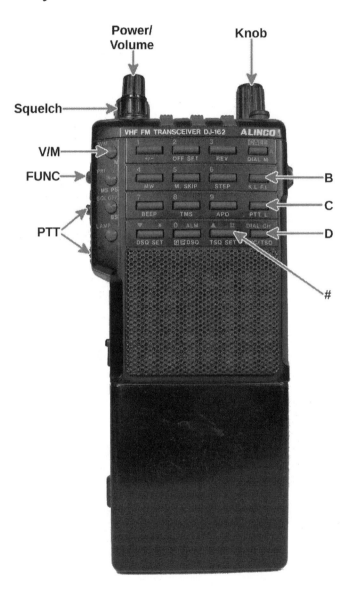

Specs

Receivers Single receiver
Receives 137.000–174.000 MHz FM
Transmits 144.000–147.995 MHz @ 5 W FM
Antenna connector BNC F on radio; needs BNC M antenna
Modes FM
Memory Channels 20
Power 9–13.8 V DC, EIAJ-02 barrel style, 4mm OD, 1.7mm ID plug, **center negative**
Model year 1992

Standard Tasks

Program frequency in the field

1. This radio doesn't allow you to overwrite memories. You will need to delete first if you want to write to a memory that has data in it. To delete: start by pressing [V/M] to enter memory mode if you aren't already there. (Screen displays **M** if you're in memory mode.) Use the knob to scroll to the memory you want to write to. If it has nothing in it, you will see a blinking **M**. If the **M** is solid, hold [FUNC] and press [4] to delete. Release [FUNC].

2. If you aren't already in VFO mode, press [V/M] to enter VFO mode. The screen will display **V**.

3. If needed, set step size: hold [FUNC] then press and release [6] then rotate the knob to select from **0 50** (5 kHz), **1** (10 kHz), **1 25** (12.5 kHz), **2** (20 kHz) and **2 50** (25 kHz). Release [FUNC] to exit.

4. Set frequency: use the keypad (144390 for 144.390 MHz).

5. If you need to change the repeater offset frequency: hold [FUNC] and press [2]. Release both buttons and use the knob to select the offset frequency in MHz. Press [V/M] to set. Note that this offset is global and applies to memories 0–16.

6. Set repeater shift: hold [FUNC] and press [1] to cycle from **-**, **+** and blank (no offset).

7. Set transmit tone: hold [FUNC] and press [#]. Release [#] but keep [FUNC] held down, and use the knob to select the tone. Release [FUNC] to set.

8. Set transmit power: slide the switch on the back of the radio. Sliding to the left (H) is high power(3 W); sliding to the right (L) is low power (0.3 W).

9. Write to a memory: hold [FUNC] then press [4].

10. Go to memory mode: press [V/M].

11. Select the memory you just wrote: use the knob.

Lock/unlock radio

The radio has two locks: a PTT lock and a key/frequency lock. To lock/unlock PTT, hold `FUNC` and press `C` (display shows PTT when locked). To lock/unlock frequency or keyboard, hold `FUNC` then press `B` to cycle through KL (key lock), FL (frequency lock) or blank (unlocked).

Check repeater input frequency

Hold `FUNC` and press `3` to switch to reverse. Hold `FUNC` and press `3` to switch back.

Change power in the field

To set transmit power, slide the switch on the back of the radio. Sliding to the left (H) is high power (3 W); sliding to the right (L) is low power (0.3 W).

Adjust volume

Use the inner part of the left knob (in front of the antenna jack) to adjust volume.

Adjust squelch

Use the outer ring of the left knob (in front of the antenna jack) to adjust squelch.

Weird Modes

Radio shows DSQ and/or transmits DTMF tones

This is not DCS as we know it. The radio has a pager mode that waits for specific DTMF tones, and sends DTMF tones on transmit. To disable, hold `FUNC` and press `0` repeatedly until DSQ disappears.

Radio doesn't transmit

This radio has a PTT lock. See the lock section to unlock.

Useful Information

This radio can have a tone squelch option installed (unit EJ-6U). If the radio has that option, you can enable tone squelch by holding `FUNC` and pressing `D`. (The display will show ENC when enabled.)

The radio has a DC power input; some batteries also have a DC power input. The battery inputs may have different ratings and/or polarities than the radio.

Memory channels 0 through 16 use the global value for offset if a shift is set. Memory channels 17 through 19 cannot have an offset.

Factory reset

Turn the radio on. Hold `FUNC` while turning the radio off and then back on again. This clears all memories and settings.

Radio Layout

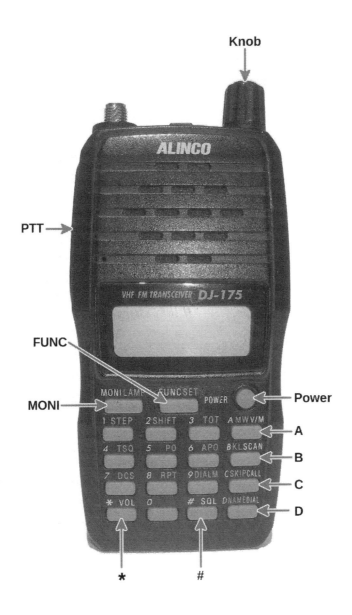

Specs

Receivers Single receiver
Receives 136.000–173.995 MHz FM
Transmits 144.000–147.995 MHz @ 5 W FM
Antenna connector SMA F on radio; needs SMA M antenna
Modes FM
Memory Channels 200
Power 12.0 V DC, No DC input on radio
Model year 2008

Standard Tasks

Program frequency in the field

1. This radio doesn't allow you to overwrite memories. You will need to delete first if you want to write to a memory that has data in it. To delete: start by pressing **A** (V/M) to enter memory mode if you aren't already there. The screen displays **M** if you're in memory mode. Press **FUNC** then rotate the knob to select the memory you want to delete. Press **A** to delete.

2. If you aren't already in VFO mode, press **A** (V/M) to enter VFO mode. The screen will show **M** when you are not in VFO mode..

3. If needed, set step size: press **FUNC** **1** (STEP) then rotate the knob to select the right **STP-** value. Press **1** to exit.

4. Set frequency: use the keypad (144390 for 144.390 MHz). If step size is greater than five, you might not have to/be able to enter the last digits.

5. Set repeater shift: press **FUNC** **2** (SHIFT). Press **2** repeatedly to cycle from **-0.600**, **+0.600** and **OST-OF** (no offset). You can adjust the shift if you need to by rotating the knob. Press **A** to set the value.

6. Set transmit tone type and value: press **FUNC** then **4** (TSQ). Pressing **4** repeatedly cycles through **T** tone encoding, **TSQ** tone encoding and decoding, and nothing (no tone). Rotate the knob to select the tone. Press **A** to save. The radio can also do DCS using **FUNC** then **7** (DCS).

7. Set transmit power: press **FUNC** **5** (PO) to cycle through **LO** (0.5 W), **MI** (2 W) and high power (5 W—nothing displayed).

8. Write to a memory: press **A** to enter memory mode.

9. Press **FUNC**. Rotate the knob to select the desired memory (0–199). Memories with data will have a solid **M**; the **M** will flash for empty memories. Make sure you're writing in an empty memory; if you pick an existing memory you will delete it instead.

10. Press **A** (MW) to write.

11. Go to memory mode: press [A] (V/M).

12. Select the memory you just wrote: use the knob.

Lock/unlock radio

Press [FUNC] [B] (KL) to lock the radio. Note that the keyboard can still send DTMF tones when locked. Press [FUNC] [B] to unlock.

Check repeater input frequency

This radio has no option to listen to the repeater input frequency. You will have to program a separate memory with the repeater input frequency to do this.

Change power in the field

To set transmit power, press [FUNC] [5] (PO). Rotate the knob to select from L0 (0.5 W), MI (2 W) and high power (5 W—nothing displayed). Press [5] to set.

Adjust volume

Press [*] (VOL). Rotate the knob to set volume from 0–20. Press [*] to set.

Adjust squelch

Press [#] (SQL). Rotate the knob to set volume from 0–10. Press [#] to set.

Weird Modes

After a reset, volume and squelch are both set to zero. You will want to change them to more reasonable values.

If the radio displays DISCHG, it is in battery refresh mode, which does a full discharge of the battery to reduce memory effect. Turn the radio off then on to leave this mode. (You can enter this mode by locking the radio, then pressing [A] [A] [B] [B] [C] [C] [D] [D].)

Useful Information

Hold [Power] for one second to turn the radio on or off.

The radio allows separate tones for encode/decode. Set one in T mode (encode) and a different one in TSQ (decode).

The radio has a "Set Mode" that allows you to set radio parameters. Enter it by holding [FUNC] for two seconds. Then you can navigate (using [FUNC] and [MONI]) through battery save BS, scan resume TIMER or BUSY, keyboard beep BEP, tone burst 1750 or another tone burst frequency, clock frequency shift SFT, busy channel lockout BCL, time after time out before you can transmit again TP, DTMF wait time DWT, DTMF pause DP, DTMF first digit time DB, and battery type BAT. Use the knob to adjust the value, and press [A] to save.

In most cases (except "Set Mode") if you wait five seconds after modifying a value, the value will be set to the new value automatically.

Factory reset

Turn the radio on while holding down **FUNC** and **A** You don't get a chance to confirm this; the radio resets as soon as you release the keys.

Settings reset

Turn the radio on while holding **FUNC** This resets everything except programmed memories. You don't get a chance to confirm this; the radio resets as soon as you release the keys.

Radio Layout

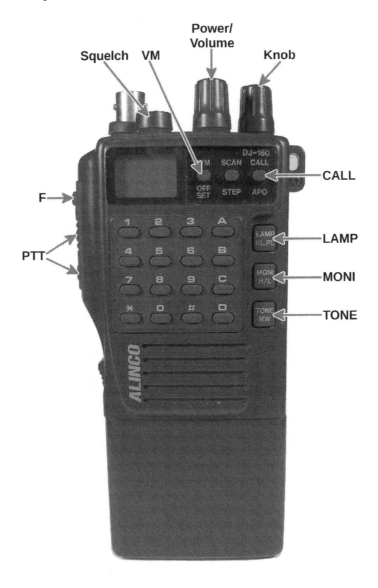

Specs

Receivers Single receiver
Receives 130.000–173.995 MHz FM
Transmits 144.000–147.995 MHz @ 5 W FM
Antenna connector BNC F on radio; needs BNC M antenna
Modes FM
Memory Channels 9 standard, memory expansion options available for 49 (EJ-14U)
    and 199 (EJ-15U)
Power 5.5–13.8 V DC, No DC input on radio
Model year 1996

Standard Tasks

Program frequency in the field

1. On this radio, you select the destination memory first. To select: make sure you're in memory mode (display shows M). Press [V/M] to get into memory mode if you're not there. Use the knob to select the memory.

2. If you aren't already in VFO mode, press [V/M]. The screen will stop showing M.

3. Set frequency: use the knob. Hold [F] while using the knob to change MHz.

4. Set transmit tone type and value: press [TONE] until display shows T (for tone on transmit) or TSQ (for tone on transmit and receive). Use knob to adjust tone value. Press [V/M] when done. You must have the EJ-17U CTCSS tone option installed for this to work.

5. Set the repeater shift and offset frequency: hold [F] and press [V/M]. Use the knob to change the offset if required. Hold [F] and press [V/M] to switch between no offset, - and +. Press [V/M] when you're done.

6. Set transmit power: hold [F] and press [MONI]. This will switch between H (high power = 5 W on 12 V battery, 2 W on 7.2 V battery) and L (low power = 0.4 W).

7. Write to a memory: hold [F] and press [TONE]. This will write to the current memory. You can press [V/M] and scroll to a different memory before writing if you need to: just press [V/M] again so you are in VFO mode when you write.

8. Go to memory mode: press [V/M].

9. Select the memory you just wrote: use the knob.

Lock/unlock radio

Hold [F] and press [LAMP]. You will cycle through KL (keyboard lock), PL (PTT lock), KLPL (both locked) and unlocked.

Check repeater input frequency

This radio has no option to listen to the repeater input frequency. You will have to program a separate memory with the repeater input frequency to do this.

Change power in the field

To set transmit power, hold [F] and press [MONI] to switch between high power (2–5 W depending on battery) and low power (0.4 W).

Adjust volume

Rotate center power/volume knob to adjust volume.

Adjust squelch

Rotate squelch knob on top left of radio.

Weird Modes

Radio displays A P

If the radio displays AP, it is in automatic power off mode. The radio will turn off if it goes 30 minutes with nothing breaking squelch. Turn the power knob off and back on again to turn the radio on again. Hold [F] and press [CALL] to enter/exit this mode.

Radio displays B

If the radio shows a B, the batteries are low. Change the batteries.

Useful Information

There are two [PTT] buttons on the left: the one marked PTT and the circular button below it. Both will transmit.

To enter extended frequency receive mode, turn radio off then turn it on while holding [LAMP]. Some versions of the radio require that you hold [F] and [LAMP] while turning it on.

This radio is available in a 70 cm version (receives 430.000–459.995 MHz FM, transmits 440.000–449.995 MHz FM @ 5 W) called the DJ-480.

Factory reset

Hold [F] while turning on radio. This clears all memories.

Radio Layout

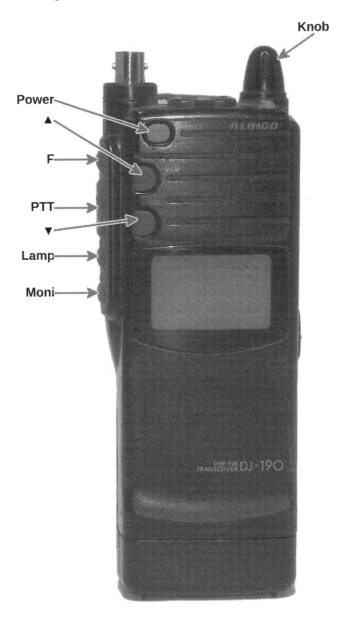

Specs

Receivers Single receiver
Receives 136.000–173.995 MHz FM
Transmits 144.000–147.995 MHz @ 5 W FM
Antenna connector BNC F on radio; needs BNC M antenna
Modes FM
Memory Channels 40
Power 4.8–13.8 V DC, No DC input on radio
Model year 1997

Standard Tasks

Program frequency in the field

1. On this radio, you select the destination memory first. To select: make sure you're in memory mode (display shows M). Hold [F] and press [▲] to get into memory mode if you're not there. Use knob to select memory to write to. The flashing M indicates an empty memory.

2. If you aren't already in VFO mode, hold [F] and press [▲]. The screen will stop showing M.

3. Set frequency: use the knob. Hold [F] while using the knob to change MHz.

4. Set transmit tone: enter menu mode by holding down [F] and pressing [MONI]. Press [▲] or [▼] until you see item 6, t NNN (where NNN is the current tone value). Use knob to adjust. Hold [F] and press [MONI] to exit.

5. Set transmit tone type: hold [F] and press [MONI]. Press [▼] to menu item 5, t-SQL. Use knob to adjust. Display will show T if tone is enabled, or T SQL if tone squelch is enabled (requires EJ-28U CTCSS decoder to be installed). Hold [F] and press [MONI] to exit.

6. If you need to change the repeater offset frequency: hold [F] and press [MONI]. Press [▼] to go to menu 4, F N.NN (where N.NN is the offset frequency). Use knob to adjust. Hold [F] and press [MONI] to exit.

7. Set repeater shift: hold [F] and press [MONI]. Press [▼] to go to menu 3, Shift. Use knob to adjust. Hold [F] and press [MONI] to exit.

8. Set transmit power: hold [F] and press [LAMP]. This will switch between high power (blank): 5 W on 12 V battery, 1.5 W on standard 4.8 V battery) and L (low power = 0.8 W).

9. Write to a memory: hold [F] and press [▼]. This will write to the current memory. You can hold [F] and press [▲] and scroll to a different memory before writing if you need to; just hold [F] and press [▲] again so you are in VFO mode when you write..

10. Go to memory mode: hold $\boxed{\text{F}}$ and press $\boxed{\blacktriangle}$.

11. Select the memory you just wrote: use the knob.

Lock/unlock radio

Hold $\boxed{\text{F}}$ and press $\boxed{\text{MONI}}$. Press $\boxed{\blacktriangle}$ or $\boxed{\blacktriangledown}$ to go to menu 1, LoC. Use the knob to scroll between blank (unlocked), KL (key lock) and FL (frequency lock). Hold $\boxed{\text{F}}$ and press $\boxed{\text{MONI}}$ to set. To disable use the same procedure.

Check repeater input frequency

This radio has no option to listen to the repeater input frequency. You will have to program a separate memory with the repeater input frequency to do this.

Change power in the field

To set transmit power, hold $\boxed{\text{F}}$ and press $\boxed{\text{LAMP}}$ to switch between high power (1.5 W–5 W depending on battery) and low power (0.8 W).

Adjust volume

Press $\boxed{\blacktriangle}$ or $\boxed{\blacktriangledown}$ to adjust volume from 0–31.

Adjust squelch

Hold $\boxed{\text{F}}$ and press $\boxed{\text{MONI}}$ to enter menu mode. Press $\boxed{\blacktriangle}$ or $\boxed{\blacktriangledown}$ to go to menu 16, SqLch. Rotate knob to adjust from 0–31. Hold $\boxed{\text{F}}$ and press $\boxed{\text{MONI}}$ to exit.

Weird Modes

Radio displays AP

If the radio displays AP, it is in automatic power off mode. The radio will turn off if it goes 30 minutes with nothing breaking squelch. Turn the radio on using the power button. To enter/exit this mode, hold $\boxed{\text{F}}$ and press $\boxed{\text{MONI}}$ to enter menu mode. Press $\boxed{\blacktriangle}$ or $\boxed{\blacktriangledown}$ to go to menu 7, APo. Use knob to adjust. (When enabled, display will show APO.) Hold $\boxed{\text{F}}$ and press $\boxed{\text{MONI}}$ to exit.

Radio displays B

If the radio shows a B, the batteries are low. Change the batteries. (The switch on the back of radio releases the battery compartment.)

Radio displays ch- 0, can't enter VFO mode

This radio has a channel-only mode. When in this mode, you can only select pre-programmed memories. Display mode. To exit (or enter) it, first lock the radio (KL or FL). Then press and release $\boxed{\text{F}}$. Then press $\boxed{\blacktriangle}$ six times, then $\boxed{\blacktriangledown}$ three times. Finally, press $\boxed{\text{MONI}}$.

Useful Information

Hold $\boxed{\textbf{Power}}$ for one second to turn the radio on or off.

Factory reset

Hold $\boxed{\textbf{F}}$ while turning on radio. This clears all memories.

Alinco DJ-500T Gen 2

Radio Layout

Specs

Receivers Single receiver, dual watch (first to break squelch wins)
Receives 136–174 MHz FM, 400–480 MHz FM
Transmits 144–148 MHz @ 5 W FM, 420–450 MHz @ 5 W FM
Antenna connector SMA F on radio; needs SMA M antenna
Modes FM
Memory Channels 200
Power 7.4 V DC, No DC input on radio
Model year 2014

Standard Tasks

Program frequency in the field

1. If you aren't already in VFO mode, press [▼]. The screen will display a number in the lower right if you're in memory mode and need to switch.

2. Set frequency: use the keypad (144390 for 144.390 MHz).

3. Set transmit tone type and value: press [A][8] to enter set menu. Use [▲][▼] to scroll to menu 01, T-CDC. Press [1] until the display shows CT (other choices are DCS and OFF). Use knob to adjust tone value. Press [D] when done. You can also adjust menu 02, R-CDC for receive tones if desired.

4. If you need to change the repeater offset frequency: press [A][8] to enter set mode. Scroll to menu 11, OFFSET using [▲][▼]. Use the knob to adjust the offset. Press [D] when done.

5. Set repeater shift: press [A][4]. Repeatedly press [4] to cycle through -, + and blank (no offset). Press [A] when done.

6. Set transmit power: press [A][9]. Repeatedly press [9] to cycle through H (high power = 5 W), M (medium power = 2.5 W) and L (low power = 1 W). Press [A] when done.

7. Write to a memory: press [A] then [▼] and the channel number will blink. Use the knob to select the desired channel. Hold [▼] until you hear two beeps.

8. Go to memory mode: press [▼].

9. Select the memory you just wrote: use the knob.

Lock/unlock radio

Press [A] then hold [#] for two seconds to lock/unlock. After unlocking, ensure you are out of function mode (F is not visible) by pressing [A] if necessary.

Check repeater input frequency

Press [A][7][A] to enter reverse mode (display shows R). Repeat to exit.

Change power in the field

To set transmit power, press [A][9]. Repeatedly press [9] to cycle through H (high power = 5 W), M (medium power = 2.5 W) and L (low power = 1 W). Press [A] when done.

Adjust volume

Rotate right power/volume knob to adjust volume.

Adjust squelch

Press [A][8] to enter the set menu. Scroll to menu 23 SQL using [▲][▼]. Use the knob to adjust squelch from 00 (open) to 09. Press [D] to exit.

Weird Modes

Can't enter VFO mode

⚠ **WARNING:** It is possible to enter channel mode from the keypad. Once you're in channel mode, there is *no way* to get back to VFO mode without a computer and cable. This mode is entered by holding down [PF2] while turning on the radio, then pressing [▲][▼] to go to menu 01, DSP. If you enable CH and press [#] you will be unable to program the VFO. This mode survives the factory reset.

Radio does not transmit

This radio has a transmit inhibit function. To disable, press [A][8] to enter the set menu. Scroll to menu 14 TX using [▲][▼]. Use the knob to switch to ON to enable transmit, or OFF to disable transmit. Press [D] to exit.

Radio button selects wrong mode

The function mode F enabled by pressing [A] is a toggle. The radio will remain in F mode until you press [A] again.

Radio doesn't switch frequencies for repeater

This radio has a talk-around mode which keeps the radio on the output frequency. To enable/disable it, press [A][8] to enter the set menu. Scroll to menu 10 TALKAR using [▲][▼]. Use the knob to switch to OFF to disable talkaround, or ON to enable it. Press [D] to exit.

Useful Information

Alinco made two generations of the DJ-500. These instructions are appropriate for the second generation with bitmap display.

Factory reset

To reset everything, hold [PF2] while turning on radio to enter special set menu. Use [▲][▼] to select menu 02, RESTOR. Use knob to select FACT?. Press [#] to reset (immediately; no prompt).

Settings reset

To reset settings but leave memories intact, hold [PF2] while turning on radio to enter special set menu. Use [▲][▼] to select menu 02, RESTOR. Use knob to select INIT?. Press [#] to reset (immediately; no prompt).

Radio Layout

Specs

Receivers Single receiver
Receives 144.000–147.995 MHz FM, 420.000–449.995 MHz FM
Transmits 144.000–147.995 MHz @ 4.5 W FM, 420.000–449.995 MHz @ 4 W FM
Antenna connector BNC F on radio; needs BNC M antenna
Modes FM
Memory Channels 100
Power 6.0–16.0 V DC, charger is 13.8 V DC, "H" barrel style jack 3.4mm OD, 1.3
    mm ID center positive. Note that DJ-596T Mk II is 6.0–15.0 V DC
Model year 2001

Standard Tasks

Program frequency in the field

1. This radio doesn't allow you to overwrite memories. You will need to delete first if you want to write to a memory that has data in it. To delete: press [A] to enter memory mode if you aren't already there. The screen displays M if you're in memory mode. Press [FUNC] then rotate the knob to select the memory you want to delete. Press [A] to delete.

2. If you aren't already in VFO mode, press [A]. The screen will not show M if you're in VFO mode.

3. If needed, change band: press [FUNC] [D].

4. If needed, set step size: press [FUNC] [1] then rotate the knob to select the right STP- value. Press [1] to exit.

5. Set frequency: use the keypad (144390 for 144.390 MHz). If step size is greater than 5, you might not have to/be able to enter the last digits.

6. Set the repeater shift and offset frequency: press [FUNC] [2]. Press [2] repeatedly to cycle from -, +, SPLIT (odd splits) and OST-OF (no offset). You can adjust the shift if you need to by rotating the knob. (Press [FUNC] to increase the adjustment factor.) Press [A] to set the value.

7. Set transmit tone type and value: press [FUNC] then [4]. Pressing [4] repeatedly cycles through T tone encoding, TSQ tone encoding and decoding, and TCS-OF (no tone). Rotate the knob to select the tone. Press [A] to save. The radio can also do DCS using [FUNC] then [7].

8. Set transmit power: press [FUNC] [5] to cycle through LO (0.8 W) and high power (2.5–5 W depending on battery—nothing displayed).

9. Write to a memory: press [A].

10. Rotate the knob to select the desired memory (0–99). Memories with data will show a solid M; the M will flash for empty memories. Make sure you're writing in an empty memory; if you pick an existing memory you will delete it instead.

11. Press [FUNC] [A] to write.

12. Go to memory mode: press [A].

13. Select the memory you just wrote: use the knob.

Lock/unlock radio

Press [FUNC] [B] to lock the radio. Note that the keyboard can still send DTMF tones when locked. Press [FUNC] [B] to unlock.

Check repeater input frequency

This radio has no option to listen to the repeater input frequency. You will have to program a separate memory with the repeater input frequency to do this.

Change power in the field

To set transmit power, press [FUNC] [5]. Rotate the knob to select from L0 (0.8 W) and high power (2.5–5 W depending on battery—nothing displayed). Press [5] to set.

Adjust volume

Press [*]. Rotate the knob to set volume from 0–20. Press [*] to set.

Adjust squelch

Press [#]. Rotate the knob to set squelch from 0–20. Press [#] to set.

Weird Modes

Volume and squelch go bad

After a reset, volume and squelch are both 0. Change them to more reasonable values.

Display shows C

You are in Call mode. Press [C] to enable/disable.

Useful Information

Hold [Power] for one second to turn the radio on or off.

The radio allows separate tones for encode/decode. Set one in T mode (encode) and a different one in TSQ (decode).

The radio has a "Set Mode" that allows you to set radio parameters. Enter it by holding [FUNC] for three seconds. Then you can navigate using [FUNC] [MONI] through battery save BS, scan resume TIMER or BUSY, keyboard beep BEP, tone burst 1750 or another tone burst frequency, busy channel lockout BCL, time after time out before you can transmit again TP, DTMF wait time DWT, DTMF pause DP, DTMF first digit time DB, theft alarm SCR (using a speaker plug where ring and ground are connected), +5 V DC on mic plug during transmit EXP, ultrasonic mosquito repellent(!) MRS and roger beep EDP. Use the knob to adjust the value, and press [A] to save.

Factory reset

Turn the radio on while holding down [FUNC]. You don't get a chance to confirm this; the radio resets as soon as you release the keys.

Radio Layout

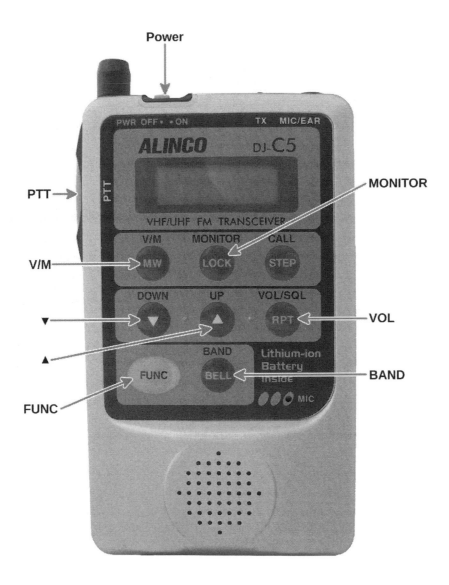

Specs

Receivers Single receiver
Receives 118.000–135.995 MHz AM, 136.000–173.995 MHz FM, 420.000–449.995
      MHz FM
Transmits 144.000–147.995 MHz @ 300 mW FM, 420–449.995 MHz @ 300 mW
      FM
Antenna connector Custom connector on radio and antenna
Modes FM
Memory Channels 50
Power 4.2 V DC (measured) to custom two-pin charge port on radio; 9 V DC, IEC
      60130-10 type C 3.8mm OD/1.4mm ID center positive to charger
Model year 1998

Standard Tasks

Program frequency in the field

1. On this radio, you select the destination memory first. To select: press `V/M` to
   get into memory mode if you're not there (display shows **M**). Press `FUNC` then
   `▲` and `▼` to scroll to the memory you want.

2. If you aren't already in VFO mode, press `V/M` to enter VFO mode. The screen
   will not show **M** in VFO mode.

3. If needed, change band: press `BAND`

4. Set frequency: press `▲`/`▼` to select desired frequency. If you press `FUNC`
   before pressing `▲` and `▼`, MHz frequency will change..

5. Set transmit tone: To set the tone, hold `FUNC` for one second, then `VOL` until
   the display shows a non-blinking **T**. Press `▲`/`▼` to select tone. Press `PTT`
   (does not transmit) to exit.

6. Set the repeater shift and offset frequency: press `FUNC` and then `VOL`. Press
   `VOL` repeatedly to change from **-**, **+** no offset. Use `▲`/`▼` to change the
   offset if required. Press `PTT` (does not transmit).

7. Power level is fixed at 300 mW.

8. Write to a memory: press `FUNC` then press `V/M`. This will write to the current
   memory..

9. Go to memory mode: press `V/M`

10. Select the memory you just wrote: use `▲`/`▼`

Lock/unlock radio

Press `FUNC` then `MONITOR`. This will lock the keyboard, but will still allow PTT
and volume/squelch adjustments.

Check repeater input frequency

Press **MONITOR** while offset is on. Press **MONITOR** again to get back to normal.

Change power in the field

This radio has a fixed 300 mW power level.

Adjust volume

Press **VOL**. Display will show `VoL.` Press **▲** **▼** to adjust volume (0–8). Press **PTT** (does not transmit) to set.

Adjust squelch

Press **VOL** twice. Display will show `SqL.` Press **▲** **▼** to adjust squelch (0–5). Press **PTT** (does not transmit) to set.

Weird Modes

Frequency not displayed

The radio can be put in a channel mode. If that happens, it will display `CH . 1`. To disable or re-enable channel mode, hold **V/M** while turning the radio on.

Useful Information

Push the **Power** switch to the right to turn the radio on. Push it to the left to turn the radio off.

The radio uses a custom antenna connector that appears to be a 2.95mm diameter threaded rod.

Custom antenna connector

Tone squelch on this radio does not provide all frequencies. You can set tone squelch from 67.0–156.7 only.

If you hold **▲** or **▼** while not adjusting a value, the radio will start to scan. Press **FUNC** to stop it.

Radios which have been modified for extended range will cycle through three values for band: 145.00, 380.00 and 445.00 (by default).

You can adjust the step size. Press **FUNC** then **CALL**. Press **▲** **▼** to adjust, then **PTT** (does not transmit) to set.

Factory reset

Hold **FUNC** and **V/M** while turning on radio, then release buttons. This clears all memories.

Radio Layout

Specs

Receivers Two independent receivers, simultaneous receive. Transmit on Main only, not Sub.
Receives 216–249.995 MHz and 902–927.995 MHz FM
Transmits 222–224.995 MHz @ 4 W FM and 902–927.995 MHz @ 1.7 W FM
Antenna connector SMA F on radio; needs SMA M antenna
Modes FM
Memory Channels 100 in each of 5 banks (Bank 0–Bank 4)
Power 9–16 V DC, IEC 60130-10 type E 3.4mm OD/1.3mm ID center positive
Model year 2011

Standard Tasks

Program frequency in the field

1. This radio doesn't allow you to overwrite memories. You will need to delete first if you want to write to a memory that has data in it. To delete: you must enable deletion if it isn't already enabled on the radio. To enable deletion, press [FUNC] [LAMP] [LAMP] then rotate Right Knob to <MEMORY>. Push [Right Knob]. Rotate Right Knob to Overwrite. Rotate Right Ring to fail-safe (allows deletion until radio is turned off) or Accepted (allows deletion always). Press [FUNC].

2. To delete a frequency, Press [VPM] until you're in memory mode. In memory mode, press [MAIN] repeatedly to go through the banks until you reach the bank you want to delete from. Use the Left Knob to select the memory in the bank that you want to delete. Press [FUNC] [CLR] [ENT] to delete that memory.

3. If you aren't already in VFO mode, press [VPM] until you're in VFO mode. The screen shows VFO.

4. If needed, set step size: press [FUNC] [7] (STEP) to adjust step size to enter the frequency. Rotate Left Knob to adjust step, then press [FUNC] again when done.

5. Set frequency: use the keypad (9270500 for 927.0500).

6. Set transmit tone type and value: press [FUNC] [5] (TONE) to enter tone set mode. Press [5] to cycle through Tx TSQ Rx OFF, Tone Squelch, Tx TSQ-Rx TSQrev, Tx TSQ Rx DCS, Tx DCS Rx OFF, Tx DCS Rx TSQ, Tx DCS-RX TSQrev, DCS, Tx OFF Rx TSQ, Tx OFF Rx TSQrev, Tx OFF Rx DCS and OFF. Rotate Left Knob to set the encoding value. Rotate Right Knob to set the decoding value (if applicable). Note that earlier versions of the firmware (before 1.10) don't have all of these options. Press [FUNC] to set.

7. This radio has automatic repeater shift hard coded for 223.91–225 MHz (-1.6 MHz) and 927–927.995 (-25 MHz). See "Useful Information" below for instructions to turn automatic repeater shift off and enable setting the repeater shift manually. To set the repeater shift once ARS has been turned off, press [FUNC] [MAIN] to enter shift mode. Press [MAIN] to cycle through -, + and

OFF (no offset). Rotate the Left Knob to select the value. Press **FUNC** when done.

8. Set transmit power: press **FUNC** **2** (PO). Use Left Knob to select from Low Power (0.4 W), Middle Power (1.0 W) and High Power (5 W on 220, 2.5 W on 900). Press **FUNC** when done.

9. Write to a memory: press **FUNC** then rotate Left Ring to select the band. (Bands 0 through 4 are "Regular" memories.) Rotate the Left Knob to select the memory slot in the band. Press **VPM** to write.

10. Go to memory mode: press **VPM**.

11. Select the memory you just wrote: use the Left Knob for main band, Right Knob for sub band. You may need to press **MAIN** to change bands.

Lock/unlock radio

Hold **FUNC** for two seconds to quick lock/unlock (black key symbol on white background). Hold **SUB** while pressing **Left Knob** three times for normal lock/unlock (white key symbol on black background). Both normal and quick lock do the same thing, just using different keys to lock/unlock.

Check repeater input frequency

Press **ENT**. Split indicator will flash to show you're in reverse mode. Press **ENT** again to return to normal mode.

Change power in the field

To set transmit power, press **FUNC** **2** (PO) then use either knob to adjust from Low Power (0.4 W), Medium Power (1.0 W) and High Power (2.5 W on 902 MHz, 4 W on 222 MHz). Press **FUNC** to set.

Adjust volume

For main (transmit) band, rotate outer ring of Left Knob to adjust volume. For sub (receive only) band, rotate outer ring of Right Knob to adjust volume.

Adjust squelch

For main (transmit) band, press **Left Knob**, then rotate Left Knob to desired squelch. Press **Left Knob** to set.

For sub (receive only) band, press **Right Knob**, then rotate Right Knob to desired squelch. Press **Right Knob** to set.

Weird Modes

The VPM button doesn't work and there's a horizontal line on the screen

You're probably in scope mode. Press [FUNC] [6] (SCOPE) to leave it.

Unable to set repeater shift between 223.91–225 MHz or between 927–927.995 MHz

Turn the automatic repeater shift off as detailed in "Useful Info."

Useful Information

To turn power on, hold [Power] for at least one second. Some users have reported the power button is stiff, so press it firmly.

The Left Knob and Right Knob on this receiver also act as buttons when pushed.

This receiver has at least two firmware versions. Version T1.00 cannot do DCS on transmit and CTCSS on receive or vice versa. Version T1.10 can.

To determine the firmware version of your receiver, hold [FUNC] for two seconds to quick lock. Then press [1] ten times to display the firmware version. Press [FUNC] to clear.

When automatic repeater shift has been turned on, you are prevented from setting the repeater shift for certain ranges. Frequencies between 223.91 and 225 MHz are set to -1.6 MHz, and frequencies between 927 and 927.995 MHz are set to -25 MHz (1.10 firmware). To disable automatic repeater shift and allow you to set the repeater shift manually, press [FUNC] [LAMP] [LAMP], then rotate Right Knob to <REPEATER>. Press [Right Knob]. Rotate Right Ring until Auto rpt. Set is OFF. Press [FUNC] to set.

Factory reset

Hold [7], [SCAN] and [FUNC] while turning radio on. This resets everything including memories.

Settings reset

Hold [FUNC] while turning radio on. This resets the radio to default settings, but preserves memories.

Radio Layout

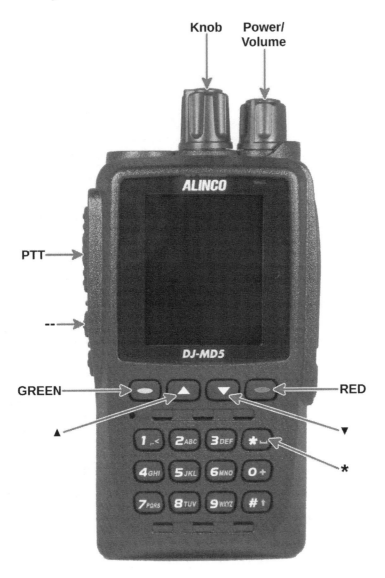

Specs

Receivers Single receiver, option of priority channel or dual watch (first to break squelch wins)
Receives 136–174 MHz FM/DMR, 400–480 MHz FM/DMR
Transmits 136–174 MHz @ 5 W FM/DMR, 400–480 MHz @ 5 W FM/DMR
Antenna connector SMA **M** on radio; needs SMA **F** antenna
Modes FM, DMR
Memory Channels 4000
Power 7.4 V DC, No DC input on radio
Model year 2018

Standard Tasks

Program frequency in the field

1. This radio requires that you program separate transmit and receive frequencies—not a receive frequency and an offset. Determine ahead of time what the transmit and receive frequencies are and have them ready.

2. To enter settings mode, press `MenuL` to bring up the menu. Press `▼` four times to get to `4 - Settings`. Press `MenuL`.

3. To set up a channel, press `▼` to select `2 Chan Set`. Press `MenuL`.

4. To set channel type, press `▼` repeatedly to select `3 CH Type`. Press `MenuL`. Use `▲`/`▼` to choose `A - Analog`, `D - Digital`, `A+D TX A` (receive analog and digital, transmit analog) or `A+D TX D` (receive analog and digital, transmit digital). Press `MenuR` to go back.

5. If you want to set both CTCSS receive and transmit tones, press `▼` to select `6 RTCDT`. Press `MenuL`. Press `▼` to select `2 CTC`, then press `MenuL`. Use `▲`/`▼` to select the tone. (You can also choose `1 DCS` if desired.) Press `MenuR` to go back.

6. To set CTCSS transmit tone only, press `▲`/`▼` to select `4 TCDT`. Press `MenuL`. Press `▼` to select `2 CTC`, then press `MenuL`. Use `▲`/`▼` to select the tone. (You can also choose `1 DCS` if desired.) Press `MenuR` to go back.

7. To set CTCSS receive tone tone only, press `▲`/`▼` to select `5 RCDT`. Press `MenuL`. Press `▼` to select `2 CTC`, then press `MenuL`. Use `▲`/`▼` to select the tone. (You can also choose `1 DCS` if desired.) Press `MenuR` to go back.

8. Set transmit power: press `▲`/`▼` to select `8 Tx Power`. Press `MenuL`. Press `▲`/`▼` to select `Small` (0.2 W), `Low` (1.0 W), `Middle` (2.5 W) or `High` (5 W). Press `MenuL` `MenuR` to set and get back to the menu.

9. To set squelch type, press `▲`/`▼` to select `9 Squelch`. Options are `SQ` (carrier squelch), `CDT` (CTCSS or DCS squelch), `Tone` (audible tone squelch), `C&T` (both CTCSS/DCS and audible tone required), `C|T` (either CTCSS/DCS or audible tone required). Press `MenuL` to set, then `MenuR` to go back.

10. To set receive frequency, press ▲ / ▼ to select 12 Rx Freq. Press **MenuL** to select, then **MenuR** to delete the frequency. Enter a new eight-digit frequency using the number keypad (144.390 MHz is 14439000). All eight digits must be entered. Press **MenuL** to confirm, then **MenuR** **MenuR** to get back to the menu.

11. To set transmit frequency, press ▲ / ▼ to select 13 Tx Freq. Press **MenuL** to select, then **MenuR** to delete the frequency. Enter a new eight-digit frequency using the number keypad (144.390 MHz is 14439000). All eight digits must be entered. Press **MenuL** to confirm, then **MenuR** **MenuR** to get back to the menu.

12. Write to a memory: press ▲ / ▼ to select 1 Store Chan. Press **MenuL** to select, then enter the channel number to write using the keypad (e.g., 1 0 for memory 10). Press **MenuL** to set. Press **MenuL** to skip the channel name. If you are prompted with Chan Exist! Replace? press **MenuL** to overwrite an existing memory.

13. Select the memory you just wrote: use the knob.

Lock/unlock radio

Hold * for one second to lock the radio. Press **MenuL** * to unlock.

Check repeater input frequency

This radio doesn't have an easy way to check the reverse frequency. You can set up a reverse memory by selecting an existing memory, then going into 4 - Settings 2 Chan Set and using 11 Reverse. Then write that to another memory.

Change power in the field

To set transmit power, you have to write the channel with a different power level. You can set up a different-powered memory by selecting an existing memory, then press **MenuL** to go into 4 - Settings 2 Chan Set. Use ▲ / ▼ to select 8 Tx Power. Press **MenuL**. Press ▲ / ▼ to select Small (0.2 W), Low (1.0 W), Middle (2.5 W) or High (5 W). Press **MenuL** **MenuR** to set and get back to the menu. Then write that to another (or the same) memory.

Adjust volume

Rotate the ring to set volume.

Adjust squelch

To set the analog squelch level, start by pressing **MenuL** and then ▲ / ▼ to choose 1 - Radio Setting. Press **MenuL** to select. Press ▼ repeatedly to reach 19 Ana SQ Level. Press **MenuL** to select. Use ▲ / ▼ to choose 0 (no squelch) or 1–5. Press **MenuL** to set, then **MenuR** repeatedly to get out of menus.

Weird Modes

Radio doesn't transmit on a memory

This radio has an option to prevent transmit on a channel. To change, press `MenuL` to go into `4 - Settings` then `2 Chan Set`. Use `▲`/`▼` to select `17 Tx Prohibit`. Press `MenuL`. Press `▲`/`▼` to select `Off` (transmit allowed) or `On` (transmit not allowed). Press `MenuL` to set. Press `MenuR` to go back, and write this back to the channel using `1 Store Chan`.

Useful Information

The radio has a VFO mode as well as a memory mode. Hold `--` to switch between modes. Unlike most radios, you don't need to start in VFO mode to program a memory. Press `--` to swap A/B.

Factory reset

Turn the radio on while holding down `PTT` and `--`. Press `MenuL` to confirm when prompted `Are you sure?` and the radio will reset.

Radio Layout

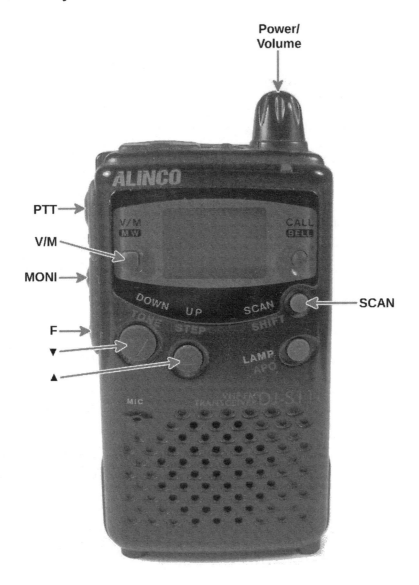

Specs

Receivers Single receiver
Receives 144.000–147.995 MHz MHz FM
Transmits 144.000–147.995 MHz @ 340 mW FM
Antenna connector Swing-out five-segment telescoping whip (built-in); no provision for alternate antenna
Modes FM
Memory Channels 20
Power 5.5 V DC, EIAJ-02 connector, 4mm OD, 1.65mm ID, center positive
Model year 1996

Standard Tasks

Program frequency in the field

1. If you aren't already in VFO mode, press [V/M] once. The screen shows M if you are not in VFO mode.

2. Set frequency: press [▲][▼] to select desired frequency. Press and release F to adjust the MHz. Press and release [F] again to adjust 100 kHz.

3. Set transmit tone: hold [F] and press [▼] (TONE). Use [▲][▼] to select tone. Hold [F] and press [▼] while in this mode to turn tone on (display shows T) or off. Press [PTT] (does not transmit) when done. This radio has tone encoding only, not tone squelch.

4. Set the repeater shift and offset frequency: hold [F] and press [SCAN] (SHIFT). Use [▲][▼] to adjust the offset (0.60 is 600 kHz). Press [SCAN] while in this mode to switch between - offset, + offset and no offset. Press [PTT] (does not transmit) when done.

5. Set transmit power: hold [PTT] and press [SCAN]. This will switch between high power (340 mW) and low power (50 mW). Adjusting the power in this way **does transmit.**

6. Write to a memory: hold [F] and press [V/M] (MW). Then use [▲] and [▼] to select the memory you want to write to (0–19).

7. Press [V/M] (MW) to write the memory.

8. Go to memory mode: press [V/M].

9. Select the memory you just wrote: use [▲][▼].

Lock/unlock radio

Hold [F] and press [MONI]. The radio will display an L. Hold [F] and press [MONI] again to unlock.

Check repeater input frequency

This radio has no option to listen to the repeater input frequency. You will have to program a separate memory with the repeater input frequency to do this.

Change power in the field

To set transmit power, hold PTT and press SCAN. This will switch between high power (340 mW) and low power (50 mW). Adjusting the power in this way **does transmit.**

Adjust volume

Rotate power/volume knob to adjust volume.

Adjust squelch

Squelch is fixed and cannot be adjusted.

Weird Modes

Radio displays OFF instead of transmitting

The radio is trying to transmit beyond the band edge. Check your offset.

Radio plays courtesy beep after transmission

Hold ▼ and turn the radio on to disable courtesy beep. Hold ▲ and turn the radio on to re-enable.

Can't enter VFO mode; display shows ch XX

You are in channel-only mode. To enter or exit this mode, hold V/M while turning the radio on.

Useful Information

Hold MONI and turn the radio on to enable/disable key beep.

The radio takes three AA batteries. Do not connect any external power to the radio while there are batteries in the radio.

Always extend the antenna when transmitting. Be careful; replacement antennas are no longer available.

Factory reset

Hold F while turning on radio. This clears all memories and settings.

Radio Layout

Specs

Receivers Single receiver with priority channel option
Receives 144.000–147.995 MHz FM, 420.000–449.995 MHz FM
Transmits 144.000–147.995 MHz @ 6 W FM, 420.000–449.995 MHz @ 6 W FM
Antenna connector SMA F on radio; needs SMA M antenna
Modes FM
Memory Channels 200
Power 4.0–15.0 V DC, charger is 13.8 V DC, "H" barrel style jack 3.4mm OD, 1.3
   mm ID center positive
Model year 2000

Standard Tasks

Program frequency in the field

1. If you aren't already in VFO mode, press [A](V/M). The screen will not show **M** in VFO mode.

2. Press [BAND].

3. If needed, set step size: press [FUNC] then [C](STEP), and then rotate the knob to select from **5.0K**, **10.0K**, **12.5K**, **20.0K**, **25.0K**, **50.0K**, **100.0K**. Press [PTT] (does not transmit) to set.

4. Set frequency: use the keypad (144390 for 144.390 MHz). If step size is greater than **5.0**, you might not have to/be able to enter the last digits..

5. Set repeater shift: press [D](RPT) repeatedly to cycle from **-**, **+**, and no display (simplex). You can adjust the shift if you need to by rotating the knob. Press [PTT] (does not transmit) to set.

6. Set transmit tone type and value: press [FUNC] then [8](T SQL). Pressing [8] repeatedly cycles through **T** tone encoding, **T SQ** tone encoding and decoding, and **OFF** (no tone). Rotate the knob to select the tone. Press [PTT] (does not transmit) to set.

7. Set transmit power: press [FUNC][6](PO). Press [6] repeatedly to cycle through **HI** (6.0 W), **L1** (1.0 W) and **L2** (0.5 W). Press [PTT] (does not transmit) to set. Note that L1 is higher power than L2.

8. Write to a memory: press [FUNC].

9. Rotate the knob to select the desired memory (0–199). Avoid other values which represent band edges and call frequencies. Memories with data will have a solid **M**; the **M** will flash for empty memories.

10. Press [A](MW) to write.

11. Go to memory mode: press [A](V/M).

12. Select the memory you just wrote: use the knob.

Lock/unlock radio

Hold **FUNC** for one second to lock the radio. Left side keys will still work when locked. Hold **FUNC** for one second to unlock.

Check repeater input frequency

Hold **0** (REV) to monitor input frequency. Release the key to return to normal operation.

Change power in the field

To set transmit power, press **FUNC** **6** (PO). Press **6** repeatedly to cycle through HI (6.0 W), L1 (1.0 W) and L2 (0.5 W). Press **PTT** (does not transmit) to set. Note that L1 is higher power than L2.

Adjust volume

Rotate the ring to set volume.

Adjust squelch

Hold **MONI** and rotate the knob to choose from L0 (open squelch) through L5. Release **MONI** to set.

Weird Modes

Can't enter frequencies

This radio has a "call mode" which prevents you from entering frequencies. When in call mode, the display will show C1 or C2. Press **A** (V/M) to go into memory mode from call mode.

Radio switches frequency every five seconds

The radio has a priority watch. If it is on, the LCD will show PRIO and will check the memory that was selected when priority was turned on. Press **FUNC** then **1** (PRIO) to turn it off/on.

Memories display as CH nn

To switch from channel display mode to frequency display mode, hold **A** while turning on power. Repeat operation to switch back.

Useful Information

Hold **Power** for one second to turn the radio on or off.

The radio allows separate tones for encode/decode. Set one in T mode (encode) and a different one in TSQ (decode).

Note that [FUNC][7] (DSQ) is DTMF squelch, not digital coded squelch.

The radio has a "Set Mode" that allows you to set radio parameters. Enter it by pressing [FUNC] then [BAND]. Then you can navigate with [*] (down) and [#] (up) through BP- (key beep), BEL (beep when DTMF squelch tone is received), APO (auto power off), BS- (battery save), DT- (DTMF squelch delay), and SP- (split). Adjust values using the knob, and press [PTT] to save.

This radio has a one-wire mod to enable multi-band receive. If that mod has been done, receive range is 76.0–999.995 MHz (cell excepted). This mod also enables extended transmit, so be careful not to program out of band. You can check by pressing [BAND] repeatedly—if you cycle through bands including 380.0 MHz and 800 MHz, the mod has been performed.

Factory reset

Turn the radio on while holding down [BAND]. You will be prompted with RESET*. Press [*] to reset.

AnyTone AT-D868UV/AT-D878UV

Radio Layout

Specs

Receivers Single receiver, dual watch (first to break squelch wins)
Receives 136–174 MHz and 400–480 MHz FM
Transmits 136–174 MHz @ 7 W FM and 400–480 @ 6 W FM
Antenna connector SMA **M** on radio; needs SMA **F** antenna
Modes FM
Memory Channels 4000
Power No DC input on radio
Model year 2017 (868), 2018 (878)

Standard Tasks

Program frequency in the field

1. This radio requires that you program separate transmit and receive frequencies—not a receive frequency and an offset. Determine ahead of time what the transmit and receive frequencies are and have them ready.

2. To create a new channel, press `MenuL` then `▲`/`▼` to select `Settings`, then press `MenuL`. Press `▼` to select `2 Channel Set` and press `MenuL`. You will be on `1 New Chan` and press `MenuL`. You will be prompted to enter a channel number; use the keypad to do so. When done, press `MenuL`. If you get the message `Same Chan!` that means you are trying to write to the channel the radio is currently tuned to—pick a different channel. Next you will be prompted for a channel name. Use the keypad to enter a name and press `MenuL` when done. If you try to write to an existing channel, you will be prompted with `Channel exists, replace?`. Press `MenuL` to replace. You will be prompted for the zone to store the channel; press `▲`/`▼` to select the right one. Press `MenuL`. Finally, you will be asked `Confirm to save`. Press `MenuL`. Press `MenuR` five times to back out of the menu.

3. Select the memory you just wrote: use the knob (otherwise you will be modifying an existing channel rather than your new one).

4. To set the channel to analog, press `MenuL` then `▲`/`▼` to select `Settings`, then press `MenuL`. Press `▼` to select `2 Channel Set` and press `MenuL`. Press `▼` repeatedly to select `3 Channel Type` and press `MenuL`. Use `▲`/`▼` to select `1 A-Analog` press `MenuL`. Press `MenuR` three times to back out of the menu.

5. Set receive frequency: press `MenuL` then `▲`/`▼` to select `Settings`, then press `MenuL`. Press `▼` to select `2 Channel Set` and press `MenuL`. Press `▼` repeatedly to select `12 Rx Freq` and press `MenuL`. Enter the value on the keypad, extending so there are five digits after the decimal point (14439000 for 144.390 MHz). Press `MenuL`. Press `MenuR` five times to back out of the menu.

6. Set transmit frequency: press `MenuL` then `▲`/`▼` to select `Settings`, then press `MenuL`. Press `▼` to select `2 Channel Set` and press `MenuL`. Press

▼ repeatedly to select 13 Tx Freq and press **MenuL**. Enter the value on the keypad, extending so there are five digits after the decimal point (14439000 for 144.390 MHz). Press **MenuL**. Press **MenuR** five times to back out of the menu.

7. Set transmit tone type and value: press **MenuL** then ▲ / ▼ to select Settings, then press **MenuL**. Press ▼ to select 2 Channel Set and press **MenuL**. Press ▼ repeatedly to select 4 TCDT (for transmit tone), 5 RCDT (for receive tone), or 6 RTCDT (to set transmit and receive tone to the same value). Press **MenuL**. Use ▲ / ▼ to select from 1 OFF (no tones), 2 CTC (CTCSS tones) or 3 DCS (DCS tones). Press **MenuL** to save. Press **MenuR** five times to back out of the menu.

8. Set transmit power: press **MenuL** then ▲ / ▼ to select Settings, then press **MenuL**. Press ▼ to select 2 Channel Set and press **MenuL**. Then press ▼ repeatedly to select 8 Tx Power and press **MenuL**. Use ▲ / ▼ to select from Power Lo (1 W), Power Mi (2.5 W), Power Hi (5 W) and Power Turbo (7 W on 2 m, 6 W on 70 cm). Press **MenuL** to save. Press **MenuR** four times to back out of the menu.

Lock/unlock radio

Hold * for 1.5 seconds to lock. Press **MenuL** then * to unlock.

Check repeater input frequency

Press **MenuL** then ▲ / ▼ to select Settings, then press **MenuL**. Next press ▼ to select 2 Channel Set and press **MenuL**. Press ▼ repeatedly to select 11 Reverse. Press **MenuL**. Use ▲ / ▼ to select from 1 REV Off or (normal) 2 REV On (reversed). Press **MenuL** to save. Press **MenuR** four times to back out of the menu.

Change power in the field

To set transmit power, press **MenuL** then ▲ / ▼ to select Settings, then press **MenuL**. Press ▼ to select 2 Channel Set and press **MenuL**. Then press ▼ repeatedly to select 8 Tx Power and press **MenuL**. Use ▲ / ▼ to select from Power Lo (1 W), Power Mi (2.5 W), Power Hi (5 W) and Power Turbo (7 W on 2 m, 6 W on 70 cm). Press **MenuL** to save. Press **MenuR** four times to back out of the menu.

Adjust volume

Rotate power/volume knob to adjust volume.

Adjust squelch

Press **MenuL** then ▲ / ▼ to select Settings, then press **MenuL**. You will be on 1 Radio Set; press **MenuL**. Press ▼ repeatedly to select 19 Ana Sq Level.

Press `MenuL` Use `▲` `▼` to select from `Ana SQ Off` or levels 1–5. Press `MenuL` to save. Press `MenuR` three times to back out of the menu.

Weird Modes

Buttons don't work as expected

The buttons on these radios are programmable. The default values are specified here. To see which buttons are mapped to which functions, press `MenuL` then `▲` `▼` to select `Settings`. Press `MenuL` You will be on `1 Radio Set`. Press `MenuL` Press `▲` `▼` to navigate to the key mappings menu items, numbers 33–42.

Can't leave channel (memory) mode

These radios are FCC Part 90 accepted, and must be unlocked for front-panel programming using a computer/radio programming cable to turn VFO mode on. Most dealers do this when selling to hams, but if the radio is locked you will need to go into the programming software (Optional Settings - Other) and switch to Amateur Mode from Professional Mode. This cannot be changed from the front panel.

Radio does not transmit

This radio has a feature to disable transmit for a memory. To enable transmit, press `MenuL` then `▲` `▼` to select `Settings`, then press `MenuL` Press `▼` to select `2 Channel Set` and press `MenuL` Press `▼` to select `17 TX Prohibit`. Press `MenuL` Use `▲` `▼` to select from `1 TX Prohibit Off` (allows transmit) or `1 TX Prohibit On` (prevents transmit). Press `MenuR` four times to back out of the menu.

Radio makes alarm noise

The radio can be programmed to have one of its buttons trigger an alarm (by default, a long press of `PF3` This does not appear to transmit. A short press of `PF3` turns off the alarm.

Knob changes frequency but doesn't select memory

This radio has a VFO mode. To enter/exit it, press `P2` You do not need to enter VFO mode to program a memory on this radio.

Useful Information

The radio can also be configured to send an alarm if it is dropped (`29 Man Down` under `Settings - Radio Set`) or if the user doesn't push a configurable button (`15 Work Alone` under `Settings - Chan Set`).

By default, a short press on `PF1` is `Volt` (battery voltage), `PF2` is `Monitor` (listen when no signal received), `PF3` is `Alarm` (transmit alarm when pressed), `P1` is `Main choose` (switch between top and bottom bands), and P2 is `VFO/MR`. Long press values may also be set.

If you get tired of pressing MenuR to back out of the menu system, just wait a few seconds and you will be returned to the root automatically.

If you delete all the channels in a zone, you will not be able to create new channels in that zone.

The Anytone AT-D868UV has an orange PF3 button. On the Anytone AT-D878UV it is blue.

This radio (with different but similar firmware) is sold as the BTech DMR-6x2.

Factory reset

To reset this radio to factory settings, hold PF1 and PTT (does not transmit) while turning the radio on. You will be prompted with `Are you sure`. Press MenuL to confirm, or MenuR to exit without initializing. After initializing, the radio will reboot and you will need to set the date. Press P1 to cycle through the fields and ▲ ▼ to change them.

⚠ **WARNING:** In at least some configurations, the AnyTone AT-D868UV/AT-D878UV may permit you to transmit on business or public safety frequencies. Make sure you are in-band when transmitting.

Radio Layout

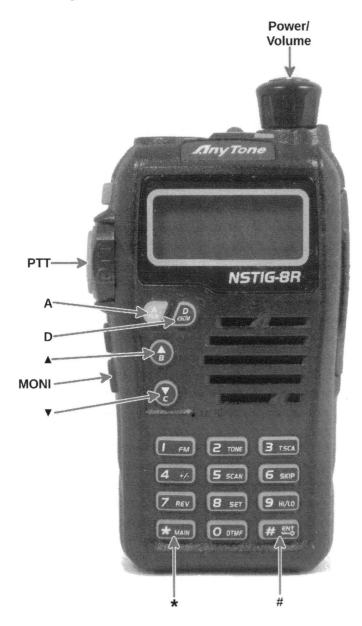

Power/
Volume

PTT

A

D

▲

MONI

▼

*

#

Specs

Receivers Single receiver, dual watch (first to break squelch wins)
Receives 136–174 MHz and 400–480 or 400–520 MHz FM
Transmits 136–174 MHz @ 5 W FM and 400–480 or 400–520 MHz @ 4 W FM
Antenna connector SMA **M** on radio; needs SMA **F** antenna
Modes FM
Memory Channels 200
Power No DC input on radio
Model year 2015

Standard Tasks

Program frequency in the field

1. If you aren't already in VFO mode, press **D**. The screen will not show a channel number next to the frequency in VFO mode.

2. Set frequency: use the keypad (144390 for 144.390 MHz).

3. Set repeater shift: press **A** then **4** (+/-). Doing this repeatedly will cycle through **-** (negative offset), **+** (positive offset) and blank (simplex).

4. If you need to change the repeater offset frequency: press **A** then **8** (SET). Use **▲**/**▼** to go to menu item 10, OFFSET. Press **A** to modify the value, then press (or hold) **▲**/**▼** to change the value. You will have to scroll to the correct value; there is no way to enter offset frequency directly. Press **D** to save and exit.

5. Set transmit tone type and value: press **A** then **8** (SET). Use **▲**/**▼** to go to menu item 01, T-CDC. Press **A** to modify the value, then press **1** to select from OFF, CT (CTCSS tone) or DCS (DCS tone). Then use **▲**/**▼** to change the value. You will have to scroll to the correct value; there is no way to enter tone frequency directly. Press **D** to save and exit.

6. Set receive tone type and value: press **A** then **8** (SET). Use **▲**/**▼** to go to menu item 02, R-CDC. Press **A** to modify the value, then press **1** to select from OFF (no receive squelch), CT (CTCSS squelch) or DCS squelch (DCS tone). Then use **▲**/**▼** to change the value. You will have to scroll to the correct value; there is no way to enter tone frequency directly. Press **D** to save and exit.

7. Set transmit power: press **A** then **9** (HI/LO) to switch between **H** (5 W on 2 m, 4 W on 70 cm) and **L** (1 W).

8. Write to a memory: press **A** then **▼**. Use **▲**/**▼** to select the memory to write (or overwrite). Press **A** and then hold **▼** for about one second to write.

9. Go to memory mode: press **D** to re-enter memory mode. You will be on the channel you just wrote.

10. Select the memory you just wrote: use ▲ ▼ to navigate to a different memory.

Lock/unlock radio

Press A then hold # for 1.5 seconds to lock/unlock.

Check repeater input frequency

Press A and then 7 (REV) to switch to reverse mode (display shows R). Press A 7 again to revert to normal.

Change power in the field

To set transmit power, press A and then 9 (HI/LO) to switch between H (5 W on 2 m, 4 W on 70 cm) and L (1 W).

Adjust volume

Rotate power/volume knob to adjust volume.

Adjust squelch

Press A then 8 (SET). Use ▲ ▼ to go to menu item 22, SQL. Press A to modify the value, then press ▲ ▼ to change the value (00–09). Press D to save and exit.

Weird Modes

Can't leave channel (memory) mode

The AnyTone NSTIG-8R is FCC Part 90 accepted, and must be unlocked for front-panel programming using a computer/radio programming cable to turn VFO mode on. This cannot be changed from the front panel.

Radio does not transmit

This radio has a transmit inhibit feature. To enable transmit, use A 8 (SET) to enter the menu, use ▲ ▼ to navigate to menu item 13, TX. Press A to modify, then use ▲ to select ON. Press D to exit.

Radio makes alarm noise

The radio can be programmed to have one of its two side buttons to trigger an alarm. This does not transmit. Turn the radio off and back on to clear it.

Useful Information

If you want to change both transmit and receive tones to the same value, use menu 3, `RT-CDC`.

Use `*` to switch between normal and inverse DCS codes when entering a DCS code.

This radio was initially sold as conforming to Part 95 as well as Part 97 and Part 90. The FCC later clarified that it was not Part 95 compliant.

Factory reset

Turn the radio off. Hold `MONI` while turning the radio on. This will enter the extended menu. Use `▲` to scroll to menu item 02, `RESTOR`. Press `A` to change, then `▲` until you have selected `FACT?`. Press `#` (ENT) to reset.

Settings reset

Turn the radio off. Hold `MONI` while turning the radio on. This will enter the extended menu. Use `▲` to scroll to menu item 02, `RESTOR`. Press `A` to change, then `▲` until you have selected `INIT?`. Press `#` (ENT) to reset.

⚠ **WARNING:** In at least some configurations, the AnyTone NSTIG-8R may permit you to transmit on business or public safety frequencies. Make sure you are in-band when transmitting.

Radio Layout

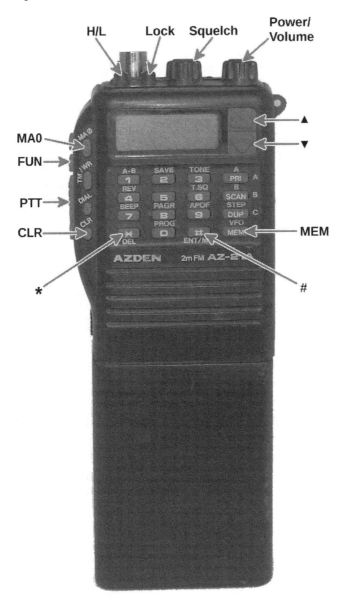

Specs

Receivers Single receiver
Receives 118.000–173.995 MHz FM
Transmits 144.000–147.995 MHz @ 5 W FM
Antenna connector BNC F on radio; needs BNC M antenna
Modes FM
Memory Channels 21
Power 12.0 V DC, EIAJ-03 barrel style, 4.7mm OD, 1.7mm ID plug, center positive
Model year 1993

Standard Tasks

Program frequency in the field

1. This radio requires that you program separate transmit and receive frequencies—not a receive frequency and an offset. Determine ahead of time what the transmit and receive frequencies are and have them ready.

2. The radio has a ten-second timeout. If you exceed this while programming, you will need to start again.

3. Hold **FUN** and **0** for one second to enter programming mode. Display will show PR.

4. On this radio, you select the destination memory first. To select: use **▲ ▼** to select memory (MA0 or 01–20). Press **#**.

5. Enter receive frequency in MHz, using ***** for the decimal point. 147*240 for 147.240 MHz. Press **#**.

6. To set receive squelch value, use **▲ ▼**. C00: means disable receive squelch. Press **#**.

7. Set frequency: use the keypad, using ***** for the decimal (147*840 for 147.840 MHz). Press **#**.

8. Set transmit tone type: use **▲ ▼**. C00: means disable receive squelch. Press **#**

9. Wait ten seconds for the memory to time out.

10. When you first program a memory, it is disabled. Press **MEM** and then very quickly enter the memory you just wrote followed by **#** (02# for memory 02).

11. To enable the memory, press **#**. The << next to the channel number will disappear. You can disable again with *****.

12. Select the memory you just wrote: use **▲ ▼**

Lock/unlock radio

Press **LOCK** to lock/unlock the radio. The button latches; up is unlocked.

Check repeater input frequency

Hold [FUN] and press [4] to go to reverse mode. Display shows REV. Repeat the sequence to go back to normal operation.

Change power in the field

Press [H/L] to switch between low (0.5 W) and high (5.0 W) power. The button latches; up is high power. Display will show LO when in low power mode. Output power cannot be changed for a particular frequency.

Adjust volume

Rotate the right knob to set volume.

Adjust squelch

Rotate the left knob to set squelch.

Weird Modes

None known.

Useful Information

The MA0 memory can be retrieved by pressing [MA0].
    To disable the keyboard beep, turn the radio on while holding [FUN].

Factory reset

To reset the radio, turn it on while holding [CLR].

Baofeng UV-100/UV-200/UV-3R

Radio Layout

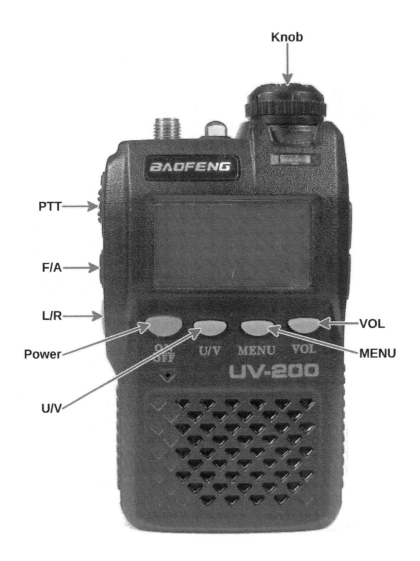

A UV-200 model

Specs

Receivers Single receiver, dual watch (first to break squelch wins)
Receives 136–174 MHz and 400–470 MHz FM
Transmits 136–174 MHz @ 2 W FM and 400–470 MHz @ 2 W FM
Antenna connector SMA **M** on radio; needs SMA **F** antenna
Modes FM
Memory Channels 99
Power No DC input on radio; 5 V DC, 2.4 mm OD, 0.70 mm ID, **center negative**
    on charger base
Model year 2011

Standard Tasks

Program frequency in the field

1. If you aren't already in VFO mode, hold [U/V] for two seconds. The channel will typically disappear, but there is a mode which makes memories look identical to frequencies.

2. If needed, set step size: press [MENU] and scroll to `09 - STEP` to set. Press [U/V] to adjust and then scroll with the knob to the correct step (5, 6.25, 10, 12.5, 20, 25). Press [MENU] to set.

3. Set frequency: use the knob. Enter UHF frequencies on the top and VHF frequencies on the bottom. Press [U/V] to switch between top and bottom. To make things quicker, you can press [F/A] before turning the knob to set MHz. Press [F/A] again to leave the MHz set mode.

4. Set transmit tone type and value: press [MENU] and scroll to `02 - TXCODE`. Press [U/V] to adjust and then scroll with the knob to the correct value (numbers with a decimal point are CTCSS; values ending in `I` or `N` are DCS). Press [MENU] to set.

5. If you need to change the repeater offset frequency: press [MENU] and scroll to `10 - OFFSET`. Press [U/V] to adjust and then scroll with the knob to the correct value. When adjusting you can press [F/A] to adjust MHz; press [F/A] again to go back to adjusting kHz. Press [MENU] to set.

6. Set repeater shift: press [MENU] and scroll to `11 - SHIFT` to set repeater shift. Press [U/V] to adjust and then scroll with the knob to the correct value (`0`, `-`, `+`). Press [MENU] to set.

7. Set transmit power: press [MENU] and scroll to `07 - POWER`. Press [U/V] to adjust and then scroll with the knob to the correct value (`HIGH`=2 W, `LOW`=1 W). Press [MENU] to set.

8. Write to a memory: press [F/A] and then [U/V]. Scroll using the knob to the desired memory. Press [U/V] again to write.

9. Go to memory mode: hold [U/V] for two seconds. Memories are available on the top side only.

10. Select the memory you just wrote: use the knob.

Lock/unlock radio

Hold [MENU] button for three seconds to lock/unlock.

Check repeater input frequency

This radio has no option to listen to the repeater input frequency. You will have to program a separate memory with the repeater input frequency to do this.

Change power in the field

To set transmit power, press [MENU] and scroll to 07 - POWER to set power level. Press [U/V] to adjust and then scroll with the knob to the correct value (HIGH=2 W, LOW=1 W). Press [MENU] to set.

Adjust volume

Press [VOL] and use knob to adjust. Press [VOL] again to set.

Adjust squelch

Press [MENU] and scroll to 03 - SQL to set squelch. Press [U/V] to adjust and then scroll with the knob to the correct value (0–9). Press [MENU] to set.

Weird Modes

Radio shows RADIO

This radio has a broadcast FM receiver in it as well. If the radio displays RADIO on the first line of the LCD and a number on the second line, hold [L/R] for two seconds to get out of this mode.

Can't set offset/direction

This radio can display frequencies instead of channel names. If you have that enabled, it's difficult to tell if you're in channel (memory) mode or frequency (VFO) mode. One sure way is to try to program an offset or shift. If it doesn't "stick" after programming, that's likely because you're in channel mode. Switch to frequency mode in order to program.

Useful Information

Hold [Power] for one second to turn the radio on or off.
     The knob locks mechanically if you push it down. Pull it up to be able to rotate.

The battery can be removed without removing the screw in the back. Flip the switch on the bottom of the radio toward the front and the back panel will slide off.

The menu times out after about ten seconds. You have to be quick when setting values.

Although the radio advertises low power as 1 W, at least some versions output less than that (about 0.1 W) on VHF. There is a hardware fix that involves replacing an 0603 47 nH coil with a 150 nH coil.

Sometimes button presses don't register on this radio. It may be easier to program if you have the keyboard beep turned on. To do this, press MENU and scroll to 05 - BEEP to set beep. Press U/V to adjust and then scroll with the knob to the correct value (OFF, ON). Press MENU to set.

Factory reset

For full reset (loses all memories and settings) hold VOL and hold Power for two seconds.

⚠ **WARNING:** In at least some configurations, the Baofeng UV-100, UV-200 and UV-3R may permit you to transmit on business or public safety frequencies. Make sure you are in-band when transmitting.

Radio Layout

Specs

Receivers Single receiver, dual watch (first to break squelch wins)
Receives 136–174 MHz and 400–480 MHz FM
Transmits 136–174 MHz @ 4 W FM and 400–480 MHz @ 4 W FM
Antenna connector SMA **M** on radio; needs SMA **F** antenna
Modes FM
Memory Channels 128
Power No DC input on radio; high capacity batteries may have an integrated 8.4 V DC barrel connector, 4.0 mm OD, 2.1 mm ID, center positive
Model year 2012

Standard Tasks

Program frequency in the field

1. This radio doesn't allow you to overwrite memories. You will need to delete first if you want to write to a memory that has data in it. To delete: press MENU 2 8 MENU (DEL-CH) *XXX* where *XXX* is the channel (001–128). Then press MENU.

2. If you aren't already in VFO mode, press VFO/MR. The channel will typically disappear, but there is a mode which makes memories look identical to frequencies. Make sure you are on the "A" side of the radio. Press A/B to switch if necessary.

3. Set frequency: use the keypad. You may need to press BAND to switch bands depending on your frequency.

4. Set transmit tone type and value: press MENU 1 3 MENU (T-CTCS) and use ▲ ▼ to select the correct CTCSS tone frequency (or off), press MENU EXIT.

5. Set repeater shift: press MENU 2 5 MENU (SFT-D) and use ▲ ▼ to select the correct repeater shift, press MENU EXIT.

6. If you need to change the repeater offset frequency: press MENU 2 6 MENU (OFFSET) and enter the value with the keypad or use ▲ ▼ to choose the correct repeater offset. Then press MENU EXIT.

7. Set transmit power: press MENU 2 MENU (TXP) and use ▲ ▼ to choose LOW (1 W) or HIGH (4 W). Then press MENU EXIT.

8. Write to a memory: press MENU 2 7 MENU (MEM-CH) and enter channel to write using the keypad *XXX* (001–128) then press MENU EXIT.

9. Go to memory mode: press VFO/MR.

10. Select the memory you just wrote: use ▲ ▼

Lock/unlock radio

Hold [#] for three seconds to lock/unlock.

Check repeater input frequency

Press [*] (display shows R). Press [*] again to revert to normal.

Change power in the field

To set transmit power, press [MENU] [2] [MENU] (TXP) and use [▲] [▼] to choose LOW (1 W) or HIGH (4 W). Then press [MENU] [EXIT].

Adjust volume

Rotate power/volume knob to adjust volume.

Adjust squelch

Press [MENU] [0] [MENU] (SQL) then use [▲] [▼] to select the desired squelch level (0–9). Then press [MENU] [EXIT].

Weird Modes

Can't leave channel (memory) mode

Some distributors ship these radios with VFO mode turned off. If that's the case, you will need to modify the programming with a computer/radio programming cable to turn VFO mode on. This cannot be changed from the front panel.

Can't set offset/direction

This radio can display frequencies instead of channel names. If you have that enabled, it's difficult to tell if you're in channel (memory) mode or frequency (VFO) mode. One sure way is to try to program an offset or shift. If it doesn't "stick" after programming, that's likely because you're in channel mode. Switch to frequency mode in order to program.

Frequency doesn't write

You must be on the A band of the radio for programming to work. Press [A/B] to switch sides.

Useful Information

If you wait too long after pressing [MENU], the radio will time out. You need to act quickly when programming.

Hold [3] while turning the radio on to see the firmware version.

Baofeng has released versions of this radio (the BF-F8 series) which look identical but claim 8 W output at high power.

This radio comes in multiple colors.

Factory reset

Press MENU 4 0 MENU (RESET) then use ▲ / ▼ to select ALL. Press MENU EXIT.

VFO reset

Press MENU 4 0 MENU (RESET) then use ▲ / ▼ to select VFO. Press MENU EXIT.

⚠ **WARNING:** In at least some configurations, the Baofeng UV-5R may permit you to transmit on business or public safety frequencies. Make sure you are in-band when transmitting.

Radio Layout

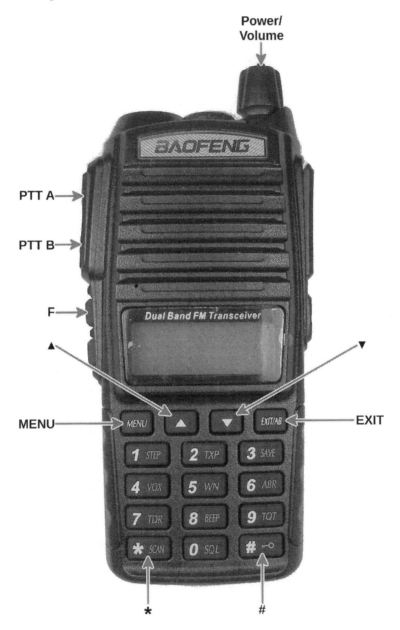

Specs

Receivers Single receiver, dual watch (first to break squelch wins)
Receives 136–174 MHz and 400–520 MHz FM
Transmits 136–174 MHz @ 5 W FM and 400–520 MHz @ 4 W FM
Antenna connector SMA **M** on radio; needs SMA **F** antenna
Modes FM
Memory Channels 128
Power No DC input on radio; high capacity batteries may have an integrated 8.4 V DC barrel connector, 4.0 mm OD, 2.1 mm ID, center positive
Model year 2014

Standard Tasks

Program frequency in the field

1. This radio doesn't allow you to overwrite memories. You will need to delete first if you want to write to a memory that has data in it. To delete: press [MENU] [2] [8] [MENU] (DEL-CH) *nnn*, where *nnn* is the channel (001–128) to delete/program. (If the display shows CH-nnn the memory has data.) Then press [MENU].

2. If you aren't already in VFO mode, turn the radio off, hold [MENU] and turn the radio on to enter VFO mode. The channel will typically disappear, but there is a mode which makes memories look identical to frequencies.

3. Set frequency: use the keypad.

4. Set transmit tone type and value: press [MENU] [1] [3] [MENU] (T-CTCS) then enter the correct tone frequency (1622 for 162.2, and 0 for OFF), press [MENU] [EXIT].

5. Set repeater shift: press [MENU] [2] [5] [MENU] (SFT-D) and use [▲] [▼] to adjust. Press [MENU] [EXIT].

6. If you need to change the repeater offset frequency: press [MENU] [2] [6] [MENU] (OFFSET) and use [▲] [▼] to adjust. Press [MENU] [EXIT].

7. Set transmit power: press [MENU] [2] [MENU] (TXP) and use [▲] [▼] to select HIGH (5 W on 2 m, 4 W on 70 cm) or LOW (1 W). Press [MENU] [EXIT].

8. Write to a memory: press [MENU] [2] [7] [MENU] (MEM-CH) and enter channel to write using the keypad *XXX* (001–128). Then press [MENU] [EXIT].

9. Go to memory mode: turn off the radio, hold [MENU] then turn it back on.

10. Select the memory you just wrote: use [▲] [▼].

Lock/unlock radio

Hold [#] for two seconds to lock/unlock.

Check repeater input frequency

Press ⟨ * ⟩ (display shows R). Press ⟨ * ⟩ again to revert to normal.

Change power in the field

To set transmit power, press ⟨ # ⟩ to switch between low power (1 W; display shows (L) and high power (5 W on 2 m, 4 W on 70 cm; display shows nothing).

Adjust volume

Rotate power/volume knob to adjust volume.

Adjust squelch

Press ⟨MENU⟩⟨ 0 ⟩⟨MENU⟩ (SQL) then adjust with ⟨▲⟩⟨▼⟩ to the desired squelch level (0–9). Press ⟨MENU⟩⟨EXIT⟩.

Weird Modes

Can't leave channel (memory) mode

The Baofeng UV-82C is FCC Part 90 accepted, and must be unlocked using a computer/radio programming cable to turn VFO mode on. This cannot be changed from the front panel. The Baofeng UV-82 is not Part 90 accepted. Both these radios can transmit on business frequencies. Be careful to make sure that you are in-band when transmitting.

Radio makes alarm noise

If you hold ⟨ F ⟩ for two seconds, the radio enters Alarm Mode. **This transmits an alarm on the currently selected frequency.** Turn the radio off and back on to clear it.

Can't set offset/direction

This radio can display frequencies instead of channel names. If you have that enabled, it's difficult to tell if you're in channel (memory) mode or frequency (VFO) mode. One sure way is to try to program an offset or shift. If it doesn't "stick" after programming, that's likely because you're in channel mode. Switch to frequency mode in order to program.

Useful Information

The Baofeng UV-82 has two PTT buttons. Normally ⟨PTT A⟩ acts as PTT for the first frequency on the display, while ⟨PTT B⟩ acts as PTT for the second frequency. On the UV-82C, ⟨PTT A⟩ can be disabled in software.

If you wait too long after pressing ⟨MENU⟩ it will time out. You need to press quickly.

Hold the [MENU] button while turning the radio on to switch from VFO mode to Memory (Channel) mode or vice versa.

Factory reset

Press [MENU] [4] [1] [MENU] (RESET) then use [▲] [▼] to select ALL. Press [MENU] [EXIT]. After reset, the 2 m offset is set to 1.600 MHz. Use menu 26 to change it.

VFO reset

Press [MENU] [4] [1] [MENU] (RESET) then use [▲] [▼] to select VFO. Press [MENU] [EXIT].

⚠ **WARNING:** In at least some configurations, the Baofeng UV-82 may permit you to transmit on business or public safety frequencies. Make sure you are in-band when transmitting.

Radio Layout

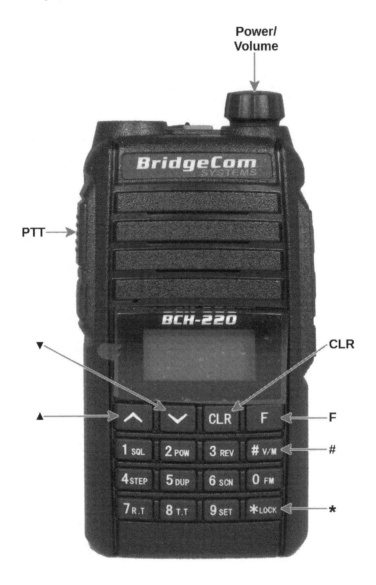

Specs

Receivers Single receiver, dual watch (first to break squelch wins)
Receives 219–260 MHz FM
Transmits 222.0–224.995 MHz @ 5 W FM
Antenna connector SMA **M** on radio; needs SMA **F** antenna
Modes FM
Memory Channels 199
Power No DC input on radio, 7.4 V DC per spec
Model year 2015

Standard Tasks

Program frequency in the field

1. If you aren't already in VFO mode, press `#` (V/M). The screen will show channel number in upper left in memory mode but not in VFO mode.

2. Set frequency: use the keypad (223500 for 223.500 MHz).

3. If you need to change the repeater offset frequency: press `F` `9` (SET) `9` `▲` to select menu 10, `OFFSET`. Press `F` then enter frequency using the numeric keypad (001600 for 1.6 MHz). Press `CLR` to exit.

4. Set repeater shift: press `F` `5` (DUP) to cycle through no offset, positive offset and negative offset. There is no indication on the front panel about the offset; you will need to watch the frequency on the display when you press `PTT` to see which way you have set. This **does transmit.**

5. Set transmit tone type and value: press `F` `8` (T.T). Press `*` to cycle through `OFF`, CTCSS tones, and DCS tones (begin with `D`). Use `▲` `▼` to select the one you want. (You can enter CTCSS tones with the keypad to get things set a little faster; 0670 for 67.0 Hz.) Then press `F`. This sets transmit tone; to set receive tone you can use `F` `7` (R.T).

6. Set transmit power: press `F` `2` (POW) then use `▲` `▼` to choose `POW HI` (high power, 5 W) or `POW LOW` (low power, 2 W). Press `F` to set.

7. Write to a memory: press `F` `*`. Then use `▲` `▼` or enter the memory to write directly from the keypad (012 is memory 12). Press `F` to save.

8. Go to memory mode: press `#` (V/M).

9. Select the memory you just wrote: use `▲` `▼` or enter a memory number on the keypad to go to the memory you just wrote.

Lock/unlock radio

Hold `*` (LOCK) for two seconds to lock/unlock. Lock icon will show when radio is locked.

Check repeater input frequency

Press [F] [3](REV) to switch to monitoring input frequency. Press [F] [3] again to revert to normal.

Change power in the field

To set transmit power, press [F] [2](POW) then use [▲][▼] to choose POW HI (high power, 5 W) or POW LOW (low power, 2 W). Press [F] to set.

Adjust volume

Rotate power/volume knob to adjust volume.

Adjust squelch

Press [F] [1](SQL) to change squelch, then use [▲][▼] to select squelch level (0–9). Press [F] to set.

Weird Modes

Display shows OFF when PTT is pressed

This means the radio is trying to transmit out of the limits of the transmit band. Adjust the offset and offset direction.

Radio makes alarm noise

This radio has an alarm mode which can be assigned to a button. If you hold the button that is assigned to the alarm for two seconds, the radio enters Alarm Mode. **This transmits an alarm on the currently selected frequency.** Turn the radio off to clear it.

Radio displays FM broadcast frequencies

This radio claims to be able to receive FM broadcast frequencies. Press [F] [0](FM) to switch back to 220 MHz.

Radio will not enter VFO mode

This radio has a channel-only mode. To enter or exit it, hold [#] while turning the radio on.

Radio locks itself after an interval

The radio has a feature to lock out the keypad after a certain amount of time. Press [F] [9](SET), then press [F]. Use [▲][▼] to select menu 02, LOCK.KEY. Press [F] and use [▲][▼] to select from MANU (no auto-lock), AT 5 (lock after five seconds), AT 10 (lock after ten seconds), AT 20 (lock after twenty seconds), AT 30 (lock after thirty seconds), ATS 5 (lock after five seconds and preserve on power off), ATS 10 (lock after ten seconds and preserve on power off), ATS 20 (lock after twenty seconds and preserve on power off), and ATS 30 (lock after thirty seconds and preserve on power off). Press [F] to set and [CLR] to exit.

Offset cannot be changed

You must be in VFO mode to change the offset. Press [#] (V/M) to switch modes.

Useful Information

The top and side buttons are programmable. By default the side is a monitor button and the top activates alarm mode.

You can enter the set menu by pressing [F] [9] (SET), and navigate it using [▲] [▼]. If you wait too long after entering set mode, it will time out. You need to press quickly. In set mode, press [F] to change the value of the currently selected parameter, and press [F] to set it. Use [CLR] to exit after setting.

Factory reset

To reset the radio, enter set mode and use [▼] until you see menu 18, RESET. Press [F]. Use [▼] to choose FULL (full factory reset). For a full reset, you will be prompted for a password—the default is 000000. Enter the password, then press [F] to reset. Note that after reset, offset reverts to 0.000 MHz and will need to be changed.

Settings reset

To reset the radio, enter set mode and use [▼] until you see menu 18, RESET. Press [F]. Use [▼] to choose either VFO (this will reset everything except memories). Then press [F] to reset. Note that after reset, offset reverts to 0.000 MHz and will need to be changed.

Harris Unity XG-100P

Radio Layout

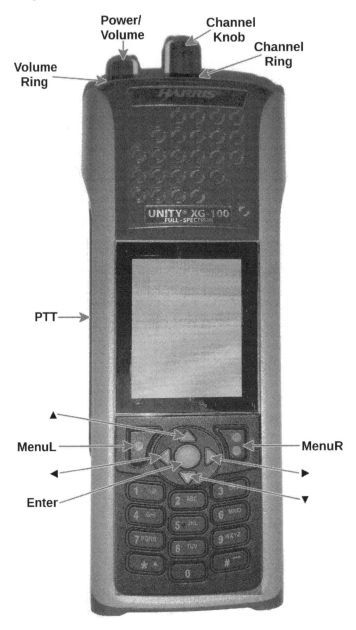

Specs

Receivers Single receiver

Receives 136–174 MHz FM, 380–520 MHz FM, 762–870 MHz FM

Transmits 136–174 MHz @ 6 W FM, 380–520 MHz @ 5 W FM, 762–870 MHz @ 3 W FM

Antenna connector SMA F on radio with special recessed connector; uses 130–870 MHz multiband antenna with SMA M connector

Modes FM, P25

Memory Channels 1000 per system, 512 systems

Power No DC input on radio; 7.5 V DC nominal, VC4000 charger input is 11–16 V DC

Model year 2010

Standard Tasks

Program frequency in the field

1. This radio has an unlock code that is required to enable front panel programming. If you don't know the unlock code, you're out of luck.

2. This radio requires that you program separate transmit and receive frequencies—not a receive frequency and an offset. Determine ahead of time what the transmit and receive frequencies are and have them ready.

3. Select the channel you want to program with the channel selector knob (and often the channel selector ring). You will overwrite the selected channel. You must use an existing analog channel; you can't overwrite a digital channel with an analog frequency.

4. Start programming by making sure you're on the main screen (while the display shows `CH INFO`) then press [MenuR].

5. Press [MenuR] (while display shows `EDIT CHAN`).

6. Enter the front panel programming unlock password by pressing the appropriate numbers on the keypad, then [Enter].

7. Set receive frequency: use [▲][▼] to scroll to `RX FREQUENCY`. Press [Enter]. Next enter the frequency using the keypad (144390 for 144.390 MHz). Press [Enter].

8. Set transmit frequency: use [▲][▼] to scroll to `TX FREQUENCY`. Press [Enter]. Next enter the frequency using the keypad (144390 for 144.390 MHz). Press [Enter].

9. Set transmit power: use [▲][▼] to scroll to `TX POWER`. Press [Enter] to switch between `LOW` and `HIGH`. Note that the values for low and high power must be set by programming software.

10. Set receive tone type and value: use [▲][▼] to select RX CHAN GUARD. Press [Enter] to bring up the menu to choose from NOISE (no tone), CTCSS (CTCSS tone) and CDCSS (DCS tones). Press [Enter] to select. Next scroll to RX TONE, press [Enter], use [▲][▼] to select the appropriate tone and press [Enter].

11. Set transmit tone type and value: use [▲][▼] to scroll to menu entry TX CHAN GUARD. Press [Enter] to bring up the menu to choose from NOISE (no tone), CTCSS (CTCSS tone) and CDCSS (DCS tones). Press [Enter] to select. Next scroll to TX TONE, press [Enter], scroll to the appropriate tone and press [Enter].

12. Press [MenuL] twice to get back to the main screen.

13. Select the memory you just wrote: use the knob.

Lock/unlock radio

The way lock is enabled can be configured through software, but many users have the volume ring configured as a lock button. Rotate it to ⊘ to lock and ◯ to unlock. An alternate method if keypad lock is turned on is to press [◄][►][▲][▼].

Check repeater input frequency

You can't switch to the repeater input frequency in the field. You will have to program a separate memory for this.

Change power in the field

To set transmit power, follow the process for editing a channel. Using [▲][▼], select TX POWER. Press [Enter] to switch between HIGH and LOW power. Depending on programming, the channel selector ring can also be configured to switch to high power. Power levels for high and low power cannot be configured in the field, but high power is usually 6 W/5 W/3 W depending on frequency.

Adjust volume

Turn the power/volume knob to adjust volume.

Adjust squelch

Squelch cannot be set in the field. Squelch level must be programmed into the radio via PC.

Weird Modes

Buttons don't work

Many of this radio's buttons and switches are programmable using a PC programmer. Depending on how the radio was configured, the buttons may or may not work, and may do different things.

Radio has open mic, flashing lights and/or sounds

The radio has an emergency mode. This transmits a signal on P25, but on analog channels it just opens the mic. The radio will also display EMERGENCY. Depending on the emergency mode programmed, there may also be exciting lights and sounds. Turn the radio off and on again to exit.

Useful Information

The volume ring may be programmed for one of many functions. They include encryption (⊘ is encrypted), TX disable (⊘ is no transmit), talkaround (⊘ is talkaround on), lock (⊘ is locked) and scan (⊘ is scanning).

This radio has multiple sets of memories (called "mission profiles"). If you make a change and then load a different mission profile, your change will be overwritten.

⚠ **WARNING:** This radio uses a recessed SMA-style connector and a multiband antenna. Some SMA antennas may not fit.

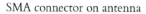

| SMA connector on antenna | SMA jack on radio |

⚠ **WARNING:** This radio is normally used in trunked systems. It will transmit its ID on such systems even if you don't push the PTT. Don't turn the radio on without an appropriate antenna when it's on a trunked channel.

No reset procedure

This radio doesn't have a reset, but you can reload an existing mission plan. To do this from the main screen, press [7] and select a plan using [▲][▼]. When you have the plan selected, press [MenuR] (when the display shows OPTIONS). Select the menu entry ACTIVATE PLAN using [▲][▼], then press [Enter] again to select. You may be asked for a PIN if the mission plan has one set.

⚠ **WARNING:** In at least some configurations, the Harris Unity XG-100p may permit you to transmit on business or public safety frequencies. Make sure you are in-band when transmitting.

Hytera AR482G

Radio Layout

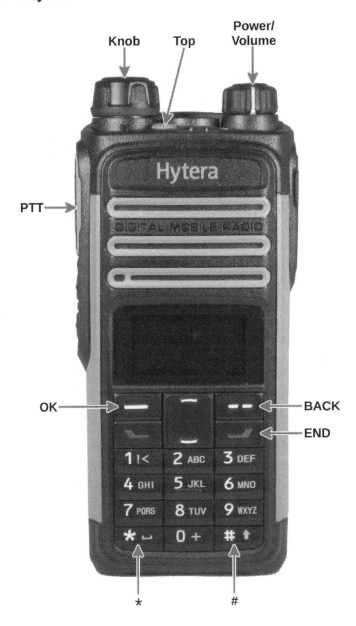

Knob   Top   Power/Volume

PTT

OK   BACK

END

*

#

Specs

Receivers Single receiver with priority channel option
Receives 420–450 MHz MHz FM/DMR
Transmits 420–450 MHz @ 4 W FM/DMR
Antenna connector SMA **M** on radio; needs SMA **F** antenna
Modes FM, DMR
Memory Channels 256
Power No DC input on radio, 7.4 V DC per spec
Model year 2018

Standard Tasks

Program frequency in the field

1. This radio requires that you program separate transmit and receive frequencies—not a receive frequency and an offset. Determine ahead of time what the transmit and receive frequencies are and have them ready.

2. Start programming by pressing `Top`. The screen shows `Manual Programming` and presents you with options for `Digital Channel` and `Analog Channel`.

3. Select `Analog Channel` by pressing `▼`. Press `OK`.

4. Edit the `Channel Alias` by pressing `OK`. Use they keypad to enter characters. `#` switches between numeric and alphabet characters. Press `*` for space, and `BACK` to erase. Press `OK` when done. Note that no two memories can have the same channel alias.

5. Set frequency: press `▼` to select `Tx Frequency`. Press `OK`.

6. Press `BACK` to erase, then enter the frequency (441150 for 441.150 MHz). Press `OK` to save.

7. Set transmit tone type: press `▼` to select `Tx CTCSS/CDCSS`. Press `OK` to save.

8. Use `▲` `▼` to select from `None` (no tone), `CTCSS` (CTCSS tones), and `CDCSS` (DCS tones). Press `OK` to save.

9. Set transmit tone: press `▼` to select `Tx CTCSS`. Press `OK` to save.

10. Press `BACK` to erase any current value, then use the keypad to enter the new tone value. Press `OK` to save.

11. Set receive tone type: press `▼` to select `Rx CTCSS/CDCSS`. Press `OK` to save.

12. Use `▲` `▼` to select from `None` (no tone), `CTCSS` (CTCSS tones), and `CDCSS` (DCS tones). Press `OK` to save.

13. Set receive tone: press `▼` to select `Tx CTCSS`. Press `OK`.

14. Press **BACK** to erase any current value, then use the keypad to enter the new tone value. Press **OK** to save.

15. Write to a memory: press **END** to go back. Then press **▼** to select `Save`. Press **OK**.

16. To select the memory to save to, use **▲** **▼** to pick the memory to overwrite. Select `1 Create Channel` to create a new channel. Press **OK** to select. If you are creating a channel, you'll be prompted one more time with `Create Channel?`. Press **OK** to confirm.

17. The channel will be added to the end of the current zone.

18. Select the memory you just wrote: rotate the left knob.

Lock/unlock radio

Press **OK** followed by ***** to lock/unlock.

Check repeater input frequency

This radio has no option to listen to the repeater input frequency. You will have to program a separate memory with the repeater input frequency to do this.

Change power in the field

To set transmit power, press **OK** then use **▲** to go to `Settings`. Press **▲** **▼** to select `Radio Set`. Press **OK**. Press **▲** **▼** to select `1 Power Level`. Press **OK**. Press **▲** **▼** to choose from `High Power` (4 W) or `Low Power` (1 W). Press **OK** to set. Press **BACK** repeatedly to get back to the main menu.

Adjust volume

Rotate the right power/volume knob to adjust the volume.

Adjust squelch

Press **OK** then use **▲** to go to `Settings`. Press **▲** **▼** to select `Radio Set`. Press **OK**. Press **▲** **▼** to select `3 SQL Level`. Press **OK**. Press **▲** **▼** to choose from `Normal`, `Open` or `Tight`. Press **OK** to set. Press **BACK** repeatedly to get back to the main menu.

Weird Modes

Radio transmits all the time

You may have enabled VOX. Turn it off by pressing **OK**, then use **▲** to go to `Settings`. Press **▲** **▼** to select `Radio Set`. Press **OK**. Press **▲** **▼** to select `VOX`. Press **OK**. Press **▲** **▼** to select `Off` and press **OK**. Press **BACK** repeatedly to back out.

Useful Information

This radio has a timeout while programming. You will need to act quickly to avoid it.

To change zones (memory banks), press `OK` then use `▼` to go to `Zones`. Press `OK`. Use `▲` `▼` to select the zone you want, then press `OK` to select.

To turn off the radio beeps, Press `OK` then use `▲` to go to `Settings`. Press `▲` `▼` to select `Radio Set`. Press `OK`. Press `▲` `▼` to select `10 Tone`. Press `OK`. Press `▲` `▼` to select `2 Keypad` and press `OK`. Use `▲` `▼` to select `Off` (or `On` if you prefer) and press `OK`. Press `BACK` repeatedly to back out.

No reset procedure

The radio does not have a way to reset it without programming a new codeplug.

Icom IC-P2AT

Radio Layout

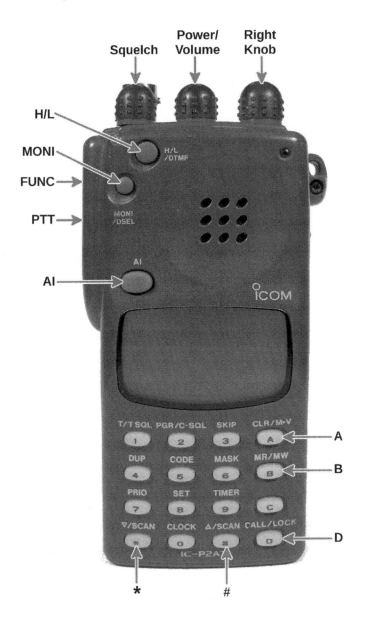

Specs

Receivers Single receiver
Receives 138–174 MHz FM
Transmits 140–150 MHz @ 5 W FM
Antenna connector BNC F on radio; needs BNC M antenna
Modes FM
Memory Channels 100
Power 6–16 V DC per spec, 13.8 V DC nominal, 3.5mm OD, 1.3mm ID barrel style
   plug, center positive
Model year 1991

Standard Tasks

Program frequency in the field

1. On this radio, you select the destination memory first. To select: press **B** (MR) to enter memory mode. Hold **FUNC** then scroll with right knob to select a memory.

2. If you aren't already in VFO mode, press **A** (CLR). The screen will not show MR in VFO mode.

3. Set frequency: enter the last four digits of frequency on numeric keypad (so 144.390 MHz is 4390). Note that if you're outside of transmit range, this will change only the last four digits of the frequency (e.g. at 160.0, pressing 4390 will change to 164.390 MHz). To get back into range quickly, hold **FUNC** and press **MONI** until there is a bar under the third digit of the frequency. Continue to hold **FUNC**, then use the right knob to adjust frequency appropriately.

4. Set transmit tone: hold **FUNC** and press **8** (SET). Press ***** (▼)/ **#** (▲) until you see TO (repeater tone). Scroll to the correct tone using the right knob. Press **A** (CLR) to select.

5. Set transmit tone type: hold **FUNC** and press **1** (T/T SQL) to activate tone (cycles through Tone, Tone Squelch with Beep, Tone Squelch, None). The display shows T when tone is active.

6. Set repeater shift: hold **FUNC** and press **4** (DUP) to select from -DUP (-)/DUP (+) /none.

7. If you need to change the repeater offset frequency: hold **FUNC** and press **8** (SET). Press ***** (▼)/ **#** (▲) until you see OW (offset). Use the right knob to scroll to the correct offset. Press **A** (CLR) to set. .60 is 600 MHz.

8. Set transmit power: hold **H/L** and use the right knob to select from high (5 W) or one of the three low levels (3.5 W, 1.5 W and 0.5 W). The S-meter indicates the level. Just pressing **H/L** switches between high and the last low level set.

9. Write to a memory: hold **FUNC** and hold **B** (MW) for two seconds.

10. Go to memory mode: press [B] (MW).

11. Select the memory you just wrote: use the right knob.

Lock/unlock radio

Hold [FUNC] then press [D] (LOCK) button to lock/unlock.
There's also a PTT lock. See "Weird Modes" for more details.

Check repeater input frequency

There is no way to listen to the repeater input frequency except to program a separate memory for it.

Change power in the field

To set transmit power, hold [H/L] and use the right knob to select from high (5 W) or one of the three low levels (3.5 W, 1.5 W and 0.5 W). The S-meter indicates the level. Just pressing [H/L] switches between high and the last low level set.

Adjust volume

Rotate center power/volume knob.

Adjust squelch

Rotate left squelch knob.

Weird Modes

PTT doesn't transmit (PTT lock)

The IC-P2AT has a PTT lock function. To disable it, press [A] to enter VFO mode. Then hold [FUNC] and press [8] (SET). Press [*] (▼)/[#] (▲) until you see PT. Use the right knob to scroll to P (allow PTT) or PL (prevent PTT). Press [A] (CLR) to set.

Not all functions are available

The IC-P2AT has a simplification feature that prevents users from entering some functions. When all features are enabled, you will see five stars at the top of the display. If you see fewer than five, you may not be able to perform some functions. To enable all functions, turn the radio off, then hold [H/L] and [AI] and turn the radio on. You will be prompted with StAr. Use the right knob to adjust to five stars, then press [PTT] (does not transmit) to set.

Useful Information

A tone module (either UT-50 or UT-51) must be installed to use CTCSS tones. The UT-50 does not provide 97.4 Hz.

Factory reset

To reset the CPU, hold **FUNC** and **A** while turning the power on. This also sets all memories and setting to default.

Icom IC-T70A

Radio Layout

Knob

Ring

PTT

SET

BAND

Power

V/M/C

H/M/L

*

Specs

Receivers Single receiver, dual watch (first to break squelch wins)
Receives 138–174 MHz FM, 400–479 MHz FM
Transmits 144–148 MHz @ 5 W FM, 420–450 MHz FM
Antenna connector SMA F on radio; needs SMA M antenna
Modes FM
Memory Channels 250
Power 10–16 V DC, 3.5mm OD, 1.3mm ID barrel style plug, center positive
Model year 2010

Standard Tasks

Program frequency in the field

1. If you aren't already in VFO mode, press [H/M/L]. The screen won't show C or WX on the main LCD, and the MR icon will not be visible when you are in VFO mode.

2. If needed, change band: press [BAND].

3. Set frequency: use the numeric keypad (144.390 MHz is 144390).

4. Set transmit tone: press [SET] then rotate knob until you see R tONE. Rotate the ring to select the tone value. You can also set tone squelch C tONE and DCS code codE. Press [H/M/L] to leave set mode.

5. If you need to change the repeater offset frequency: press [SET] then rotate knob until you see OFFSEt. Rotate ring to select the offset value. Press [H/M/L] to leave set mode.

6. Set transmit tone type: repeatedly hold for one second and then release [0] to cycle through T (tone), ((.)) T SQL (tone squelch with beep), T SQL (tone squelch), ((.))DTCS (DCS with beep), DTCS (DCS), T SQL-R (squelch if a particular tone *is* received), DTCS -R (squelch if a particular code *is* received), and blank (no tone/code).

7. Set repeater shift: repeatedly hold for one second and then release [*] to cycle through DUP - (-), DUP (+) and blank (simplex).

8. Set transmit power: press [H/M/L] repeatedly to cycle through high (5 W), medium (2.5 W, radio shows M) and low (0.5 W, radio shows L).

9. Write to a memory: hold [V/M/C] for one second and release. Use knob to scroll to a memory (0–249). Stay away from channels with A or b in them; those are scan edge channels. Hold [V/M/C] for one second and release to write.

10. Go to memory mode: press [V/M/C].

11. Select the memory you just wrote: use the knob.

Lock/unlock radio

Hold [SET] for one second to lock/unlock.
There's also a PTT lock. See "Weird Modes" for more details.

Check repeater input frequency

Hold [BAND]. Radio will enter monitor mode and switch to reverse frequency. Release [BAND] to go back to normal.

Change power in the field

To set transmit power, press [H/M/L] repeatedly to cycle through high (5 W), medium (2.5 W, radio shows `M`) and low (0.5 W, radio shows `L`).

Adjust volume

Rotate ring; volume is from 0–24.

Adjust squelch

Hold [BAND] and rotate knob. Settings are `OPEN`, `Auto`, then `LEVEL1` through `LEVEL9`.

Weird Modes

Radio displays `Hot`

The radio has protection from overheating. It will first fold power back to medium from high, and then prevent transmission entirely. Cool the radio down.

PTT doesn't transmit (PTT lock)

The IC-T70A has a PTT lock function. To disable it, hold press [SET] while turning on the radio. Rotate the knob until you see `Ptt Lk`. Rotate the ring until you see `OFF PtL`. Press [Power]. To turn it back on, repeat the same steps using the ring to select `ON PtL`.

Not all frequencies are available in memory mode

The IC-T70A can have memory banks configured. If it does, you can turn them off. While in memory mode, press [BAND]. Next rotate the knob until you don't see `bAnk`. Then press [BAND] again.

Useful Information

Hold [Power] for one second to turn the radio on or off.
Rotating the ring on this radio can sometimes cause the knob to rotate as well, and vice versa. Hold one down carefully while adjusting the other.

The IC-T70A speaker mic has an input impedance of 2.2 kΩ. Some third-party speaker mics may not be properly matched.

The display will show two different kinds of WX. There's a WX icon which is visible when weather alert is turned on. If you're actually on the weather band, you'll see WX-01 through WX-10 on the main LCD display.

Factory reset

To reset everything including memories, hold [SET], [BAND] and [H/M/L] while turning the radio on. The radio will reset immediately with no prompt.

Settings reset

To reset most settings but leave memories intact (partial reset), hold [SET] while turning the radio on. The radio will reset immediately with no prompt.

Icom IC-V8

Radio Layout

Volume

Power

PTT

▲

SQL

▼

A

C

D

#

Specs

Receivers Single receiver
Receives 136–174 MHz FM
Transmits 144–148 MHz @ 5.5 W FM
Antenna connector BNC F on radio; needs BNC M antenna
Modes FM
Memory Channels 100
Power No DC input on radio; 6–10.3 V DC per spec
Model year 2001

Standard Tasks

Program frequency in the field

1. If you aren't already in VFO mode, press **D** (CLR). The screen will show MR if you aren't.

2. Set frequency: use the numeric keypad (144390 for 144.390 MHz).

3. Set transmit tone: press **A** (FUNC) **8** (SET) and push **▲**/**▼** until you see **rt** (repeater tone). Scroll to the correct tone using the volume knob. Press **#** (ENT) to select.

4. Set transmit tone type: press **A** (FUNC) **1** (TONE) repeatedly. Different icons will appear on the display. They are: ♪ (transmit CTCSS tone), ◁ (CTCSS tone squelch), Ⓓ (DCS tone and squelch) and blank (no tone).

5. Set repeater shift: repeatedly press **A** (FUNC) **4** (DUP) to select from −/+/none.

6. If you need to change the repeater offset frequency: press **A** (FUNC) **8** (SET) and push **▲**/**▼** until you see ± (offset). Scroll to the correct tone using the volume knob. Press **#** (ENT) to select. 0.60 is 600 MHz.

7. Set transmit power: press **A** (FUNC) then **9** (HI/LOW) to select between 5.5 W (shows nothing on right) and 0.5 W (shows L on right).

8. Write to a memory: press **A** (FUNC) **C** (MR). MR will flash.

9. Use **▲**/**▼** to choose a memory.

10. Press **A** (FUNC) then hold **C** (MR) for one second to write.

11. Go to memory mode: press **C** (MR).

12. Select the memory you just wrote: use **▲**/**▼**. Note that **D** (CLR) goes back to VFO mode; after pressing it you'll need to press **C** (MR) again.

Lock/unlock radio

Press **A** (FUNC) then hold **#** (ENT) button for one second to lock/unlock.

Check repeater input frequency

Press ⎡D⎤(CLR) if you're not already in VFO mode, then ⎡A⎤(FUNC)⎡8⎤(SET) and push
⎡▲⎤⎡▼⎤until you see REV.OF (reverse mode off). Use the volume knob to change to
REV.ON, then press ⎡#⎤(ENT) to select. In reverse mode the + or - symbol will blink.
Repeat procedure to switch back to REV.OF.

Change power in the field

To set transmit power, press ⎡A⎤(FUNC) then ⎡9⎤(HI/LOW) to select between 5.5 W
(shows nothing on right) and 0.5 W (shows L on right).

Adjust volume

Turn the volume knob to adjust volume.

Adjust squelch

Hold ⎡SQL⎤ button and push ⎡▲⎤⎡▼⎤

Weird Modes

Volume knob changes frequency rather than volume

The IC-V8 has an automobile mode which switches the functions of the volume knob
and the ⎡▲⎤⎡▼⎤buttons. If you're in it, turn the radio off. Then hold both ⎡▲⎤and
⎡▼⎤while turning the radio on. Scroll with the ⎡▲⎤⎡▼⎤buttons get to toP.dl (knob
is for tuning). Then use the volume knob to change it to toP.VO (knob is for volume).
Press ⎡#⎤(ENT) to select.

Radio doesn't transmit

The IC-V8 can be set to prevent transmission. Use ⎡D⎤(CLR) if you're not already in
VFO mode, then ⎡A⎤(FUNC)⎡8⎤(SET) and scroll with ⎡▲⎤⎡▼⎤to tX. Use volume to
adjust between tX.ON (allows transmit) and tX.OF (prevents transmit). Press ⎡#⎤(ENT)
to set.

Function key works strangely

The behavior of the ⎡A⎤(FUNC) key can be changed. Use ⎡D⎤(CLR) if you're not already
in VFO mode, then ⎡A⎤(FUNC)⎡8⎤(SET) to change. F0.AT is the default, meaning
function mode disappears zero seconds after using a function. It can also be set to F1.AT,
F2.AT, F3.AT (function mode disappears one, two, three seconds after using a function)
or F.m (you have to hit ⎡A⎤(FUNC) to turn off function mode).

Can't get out of memory mode

The IC-V8 has a channel mode which prevents direct frequency entry. In this mode,
channels will display as CH 27. To get out of it, turn the radio off. Then hold both ⎡▲⎤

and $\boxed{\blacktriangledown}$ while turning the radio on. Scroll with the $\boxed{\blacktriangle}$ $\boxed{\blacktriangledown}$ buttons get to `dsp.CH` (display channel mode). Use the volume knob to change it to `dsp.FR` (frequency) or `dsp.Nm` (name). Press $\boxed{\text{\#}}$ (ENT) to select.

Useful Information

Hold $\boxed{\textbf{Power}}$ for one second to turn the radio on or off.

⚠ **WARNING:** According to an article from Icom (see the Icom website at `https://www.icomjapan.com/explore/genuine_info`) the IC-V8 has been counterfeited, and the counterfeit is being sold at online auction sites, among other places. The counterfeit radios can be identified by a number of ways, the most obvious being that the $\boxed{\text{2}}$ reads "VOX" instead of "P.BEEP," and the speaker mic jacks have white labels instead of no labels. The menus for these counterfeit devices are different from the Icom menus.

Tone squelch can use a different frequency than CTCSS tone. Use the same procedure to adjust tone squelch as you do for tone, but scroll to `Ct` for tone squelch and `dt` for DCS.

Factory reset

To reset the CPU, hold $\boxed{\text{D}}$ (CLR) and $\boxed{\textbf{SQL}}$ while pushing the $\boxed{\textbf{Power}}$ button. This also clears all memories and settings.

Icom IC-V80 HD

Radio Layout

Specs

Receivers Single receiver with priority channel option
Receives 136–174 MHz FM
Transmits 144–148 MHz @ 5.5 W FM
Antenna connector BNC F on radio; needs BNC M antenna
Modes FM
Memory Channels 200
Power No DC input on radio; 7.2–9 V DC based on available batteries
Model year 2010

Standard Tasks

Program frequency in the field

1. If you aren't already in VFO mode, press V/M/C (VFO/MR/CALL). The screen will show MR, CO or WX if you aren't.

2. Set frequency: use the numeric keypad (144390 for 144.390 MHz).

3. Set transmit tone: press * (FUNC) 8 (SET) and push ▲ / ▼ until you see rt (repeater tone). Scroll to the correct tone using the knob. Press #ENT to set. Use dt for DCS.

4. Set transmit tone type: press * (FUNC) 1 (TONE) repeatedly. Different icons will appear on the display. They are: ♪ (transmit CTCSS tone), ◁)) (CTCSS tone squelch with "pocket beep" paging), ◁ (CTCSS tone squelch), D)) (DCS squelch with "pocket beep" paging), D (DCS squelch), and blank (no tone).

5. Set repeater shift: repeatedly press * (FUNC) 4 (DUP) to select from -/+/none.

6. If you need to change the repeater offset frequency: press * (FUNC) 8 (SET) to adjust. Use ▲ / ▼ to get to ±. Use the knob to adjust the value, and press #ENT to set.

7. Set transmit power: press * (FUNC) 9 (H/M/L) to select from 5.5 W (shows nothing on screen), 2.5 W (shows M on screen) and 0.5 W (shows L on screen).

8. Write to a memory: press * (FUNC) then hold V/M/C (VFO/MR/CALL).

9. Use ▲ / ▼ to choose a memory.

10. Hold V/M/C (VFO/MR/CALL) to write.

11. Go to memory mode: press V/M/C (VFO/MR/CALL).

12. Select the memory you just wrote: use ▲ / ▼

Lock/unlock radio

Press * (FUNC) then hold #ENT button for one second to lock/unlock.

Check repeater input frequency

Press [*] (FUNC) [8] (SET) and push [▲]/[▼] until you see REV.OF (reverse mode off). Use the knob to change to REV.ON, then press [#ENT] to select. In reverse mode the + or − symbol will blink. Repeat procedure to switch back to REV.OF. You can also hold [MONI] to temporarily check the input frequency of a repeater memory.

Change power in the field

To set transmit power, press [*] (FUNC) [9] (H/M/L) to select from 5.5 W (shows nothing on screen), 2.5 W (shows M on screen) and 0.5 W (shows L on screen).

Adjust volume

Turn the knob to adjust volume.

Adjust squelch

Hold [MONI] button and push [▲]/[▼] to go from level 0–10.

Weird Modes

Knob changes frequency rather than volume

This radio can be configured so the knob changes frequency instead of volume. To adjust, turn the radio off. Then turn it on while holding [▲] and [▼]. Press [▲]/[▼] to scroll to tOP.. Use the knob to select from tOP.VO (volume) or tOP.dI (frequency). Press [#ENT] to select.

Radio doesn't transmit

This radio has a transmit inhibit function which can be stored with individual memories. To disable, press [*] (FUNC) [8] (SET) to enter set mode and and push [▲]/[▼] until you see tX OF. Scroll using the knob to tX ON. Press [#ENT] to set.

Can't get out of memory mode

This radio can be configured to use a memory-only mode. To exit, turn the radio off. Then turn it on while holding [▲] and [▼]. Press [▲]/[▼] to scroll to dSP.Ch. Use the knob to select either dSP.FR (display frequency) or dSP.Nm (display name). Press [#ENT] to select.

Useful Information

Hold [Power] for one second to turn on/off.

Tone squelch can use a different frequency than CTCSS tone. Use the same procedure to adjust tone squelch as you do for tone, but scroll to Ct for tone squelch and dt for DCS.

Factory reset

To reset everything, turn the radio off. Then hold **V/M/C** (VFO/MR/CALL) and **MONI** and while turning the radio on. The display will show `CLEAR` and the radio will be completely cleared.

Settings reset

To reset settings, turn the radio off. Hold **V/M/C** (VFO/MR/CALL) while turning the radio on. Nothing obvious will happen but the settings will be reset.

Icom IC-V86/IC-U86

Radio Layout

Knob

PTT

MONI

Power

▲

▼

V/M/C

*

#ENT

Specs

Receivers Single receiver with priority channel option
Receives V86: 136–174 MHz FM, U86: 400–472 MHz
Transmits V86: 144–148 MHz @ 7 W FM, U86: 430–450 MHz @ 5.5 W
Antenna connector BNC F on radio; needs BNC M antenna
Modes FM
Memory Channels 200
Power No DC input on radio; 7.5 V nominal
Model year 2019

Standard Tasks

Program frequency in the field

1. If you aren't already in VFO mode, press **V/M/C** (VFO/MR/CALL). The screen will display **VFO**.

2. Set frequency: enter the frequency on the keypad (144390 for 144.390 MHz).

3. Set transmit tone type: press ***** (FUNC) **1** (TONE) repeatedly. Different icons will appear on the display. They are: ♪ (transmit CTCSS tone), ◁)) (CTCSS tone squelch with "pocket beep" paging), ◁ (CTCSS tone squelch), D♪ (transmit DCS tones), D)) (DCS squelch with "pocket beep" paging), D (DCS tone squelch), and blank (no tone).

4. Set transmit tone: press ***** (FUNC) then **8** (SET). Use **▲**/**▼** to select the value to modify, and use the knob to change the value. Values to modify are **rt** CTCSS tone to send, **Ct** CTCSS squelch tone, **dt** DCS tone, **dtP** DCS polarity (up/down, **N** is normal, **R** is reverse). Press **#ENT** to set.

5. Set repeater shift: repeatedly press ***** (FUNC) then **4** (DUP) to cycle through **-** (negative offset), **+** (positive offset) and blank (simplex). Press **#ENT** to set.

6. If you need to change the repeater offset frequency: press ***** (FUNC) then **8** (SET). Use **▲**/**▼** to select **±**, and use the knob to change the value. Press **#ENT** to set.

7. Set transmit power: repeatedly press ***** (FUNC) then **9** (H/M/L) to cycle through **H** (high power, 5 W on the V86, 4 W on the U86), **M** (medium power, 2.5 W on the V86, 2 W on the U86), **L** (low power, 0.5 W on the V86, 0.5 W on the U86) and **EXH** (extra-high power if enabled, 7 W on the V86, 5.5 W on the U86).

8. Write to a memory: press ***** (FUNC) **V/M/C** (VFO/MR/CALL).

9. Use **▲**/**▼** to choose a memory.

10. Press ***** (FUNC) then hold **V/M/C** (VFO/MR/CALL) to write to that memory.

11. Go to memory mode: press **V/M/C** (VFO/MR/CALL).

12. Select the memory you just wrote: use **▲**/**▼**

Lock/unlock radio

Press [*](FUNC) then [#ENT] to lock/unlock.

Check repeater input frequency

Press [*](FUNC) then [8](SET). Use [▲][▼] to choose REV.OF. Use the knob to change that to REV.ON. Press [#ENT] to set. In reverse mode the offset indicator will blink. Repeat the procedure, usign the knob to change to REV.OF to get back to normal.

Change power in the field

To set transmit power, repeatedly press [*](FUNC) then [9](H/M/L) to cycle through H (high power, 5 W on the V86, 4 W on the U86), M (medium power, 2.5 W on the V86, 2 W on the U86), L (low power, 0.5 W on the V86, 0.5 W on the U86) and EXH (extra-high power if enabled, 7 W on the V86, 5.5 W on the U86).

Adjust volume

Rotate the knob to adjust volume (0–24).

Adjust squelch

Hold [MONI] then use [▲][▼] to change the squelch level (0–10). Release to set.

Weird Modes

Radio doesn't transmit

This radio has a transmit inhibit setting. Press [*](FUNC) then [8](SET) and use [▲][▼] to get to tX .OF. Use the knob to change it to TX. ON. Press [#ENT] to set.

Useful Information

Hold [Power] for about two seconds to turn the radio off or on.

To enter the extra-high power mode, you must first enable it in initial settings. To do this, turn off the radio. Hold [▲] and [▼] and while keeping both held down, turn the radio on. Use [▲][▼] to go to EXH. Then use the knob to change the setting: EXH.ON enables extra-high-power mode, EXH.OFF disables it.

Factory reset

To reset everything, turn the radio off. Then hold [MONI] and [V/M/C] (VFO/MR/CALL) while turning the radio on. You will see CLEAR and the radio will reset itself (there is no prompt).

Icom IC-W32A

Radio Layout

Specs

Receivers Two independent receivers, simultaneous receive
Receives 118–135.995 MHz AM, 136–174 MHz FM, 440–470 MHz FM
Transmits 144–148 MHz @ 5 W FM, 440–450 MHz @ 5 W FM
Antenna connector BNC F on radio; needs BNC M antenna
Modes FM
Memory Channels 100 VHF, 100 UHF
Power 4.5–16 V DC per spec, 13.5 V DC nominal, 3.5mm OD/1.3mm ID center
positive (on battery—different batteries may have different connectors)
Model year 1996

Standard Tasks

Program frequency in the field

1. If needed, change band: press [MAIN].

2. If you aren't already in VFO mode, press [VFO]. The screen will show M, C or a memory name if you aren't in VFO mode.

3. Set frequency: use the numeric keypad (144390 for 144.390 MHz).

4. Hold [H/L] for two seconds to enter set mode.

5. Set transmit tone: hold [H/L] for two seconds to enter set mode. Press [H/L] or [TONE] until you see RT. Use the appropriate knob (left or right) to select the desired tone. You can also use [H/L] and [TONE] to move to CT to set a different tone for tone squelch if desired. Press [VFO] to exit set mode.

6. If you need to change the repeater offset frequency: hold [H/L] for two seconds to enter set mode. Press [H/L] or [TONE] until you see OW. Use the appropriate knob to adjust offset (note that offset is global for the selected band). Press [VFO] to exit set mode.

7. Set repeater shift: hold [TONE] for two seconds. Repeat to cycle through -DUP (negative), DUP (positive) and blank (no shift).

8. Set transmit tone type: press [TONE] repeatedly to enable tone encoder. Display will cycle through T (tone on transmit, no tone on receive), T SQL(.) (tone page), T SQL (tone on transmit, tone on receive).

9. Set transmit power: press [H/L] to select between high power (5 W) and low power (0.5 W, display shows LOW).

10. Write to a memory: press [S.MW].

11. Use the appropriate knob to select the correct memory.

12. Hold [S.MW] for two seconds to write.

13. Go to memory mode: press [MR].

14. Select the memory you just wrote: use the appropriate knob.

Lock/unlock radio

Hold CALL for two seconds to lock/unlock. Lock just prevents frequency change.

Check repeater input frequency

There is no way to listen to the repeater input frequency except to program a separate memory for it.

Change power in the field

To set transmit power, press H/L to select between high power (5 W) and low power (0.5 W, display shows LOW).

Adjust volume

Rotate left volume ring for left band volume; rotate right volume ring for right band volume.

Adjust squelch

Hold SQL while rotating left knob or right knob. Values are AT (auto squelch), OPN (open) and SQ1 through SQ8.

Weird Modes

Radio shows CH:01 or 118:00

The radio can tune to NOAA weather channels (CH:01 through CH:10) and the air band (118–135.995 MHz). Press BAND repeatedly until you see the 2 m and 70 cm bands.

Radio shows HCH:nn or LCH:nn

The radio can display either channel names or channel numbers. Push M.N to switch between them.

Display is hard to read

The LCD contrast is adjustable. To change, hold H/L while turning the radio on. Then press TONE nine times. If you can see the display, it will show LC. Rotate the right knob to select contrast. Turn the radio off to save.

Display shows LOW V or OVER V when turned on

LOW V means the unit has less than 4.5 V available. OVER V means the device is getting more than 16 V.

Useful Information

Hold [Power] for two seconds to turn the radio on or off.

The buttons [H/L] and [TONE] are sometimes used to scroll up/down.

If you write into a memory that previously had a channel name, that name will be retained. This may be confusing.

The device has an initial setup menu that is entered by holding [H/L] while turning on the device. From there, you can adjust MS (handheld mic mode), AO (auto power off), LI (backlight), BE (keyboard beep), AR (auto repeater shift), PS (power saver), VO (battery voltage display), DT (DTMF speed), LC (LCD contrast) and CB (crossband repeat mode). Use [H/L] [TONE] to scroll through them, and set the value with either knob. Turn the power off again to save the values.

This radio has a built-in manual. Hold [L/G] while pressing any button to learn more about that button. Press any button to exit.

This radio can crossband repeat. Set up is done using the initial setup menu. Modes are SEMI CB (mute sub-band when main band is transmitting) and FULL CB (two-way). To turn on crossband repeat: set the main and sub band frequencies, then lock the radio by holding [CALL]. Then turn the radio off. Finally hold [MAIN], [BAND] and [SQL] while turning the radio on. The lock icon will flash. Unlock (by holding [CALL]) to turn off crossband repeat.

The radio can be put in an extended receive mode by holding down [MAIN], [BAND], [CALL] and [SQL] while turning the radio on.

Factory reset

To reset everything (including memories), hold [SQL], [VFO] and [MR] while turning the power on.

Settings reset

To reset the CPU and settings only (but not memories), turn the radio on while holding down [VFO].

Radio Layout

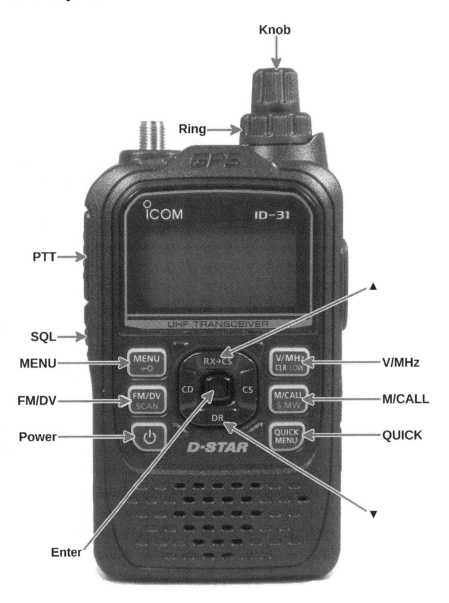

Specs

Receivers Single receiver with priority channel option
Receives 400–479 MHz FM, DSTAR
Transmits 420–450 MHz FM, DSTAR @ 5 W FM, DSTAR
Antenna connector SMA F on radio; needs SMA M antenna
Modes FM, DSTAR
Memory Channels 500
Power 10–16 V DC (charger is 12 V DC), 3.5mm OD/1.3mm ID center positive
Model year 2012

Standard Tasks

Program frequency in the field

1. If you aren't already in VFO mode, press [V/MHz]. The screen will show MR if you aren't.

2. If needed, set step size: press [QUICK] then use [▲]/[▼] to select TS. Press [Enter]. Use [▲]/[▼] to select the step size you want, then press [Enter].

3. Set mode: press [FM/DV] to cycle through FM (wide FM), FM-N (narrow FM) and DV (DSTAR).

4. Set frequency: use the knob. Press [V/MHz] to cycle through 1 MHz adjustments or normal step size.

5. Set repeater shift: press [QUICK]. Use [▲]/[▼] to navigate to DUP. Press [Enter]. Use [▲]/[▼] to select DUP- (negative offset), DUP+ (positive offset) or OFF (simplex). Press [Enter] to set and exit the menu.

6. If you need to change the repeater offset frequency: press [MENU] then use [▲]/[▼] to navigate to DUP/TONE. Press [Enter]. Use [▲]/[▼] to select Offset Freq. Press [Enter]. Use the knob to adjust offset. Pressing [V/MHz] will cycle through 1 MHz, 10 MHz or step size. Press [MENU] to exit the menu.

7. Set transmit tone: (for transmit) press [MENU]. Use [▲]/[▼] to navigate to DUP/TONE. Press [Enter]. Use [▲]/[▼] to select Repeater Tone. Press [Enter]. Use knob to adjust. Press [MENU] to exit the menu.

8. Set transmit tone: (for receive) press [MENU]. Use [▲]/[▼] to navigate to DUP/TONE. Press [Enter]. Use [▲]/[▼] to select TSQL Freq. Press [Enter]. Use knob to adjust. Press [MENU] to exit the menu.

9. Set transmit tone type: press [QUICK]. Use [▲]/[▼] to navigate to TONE. Press [Enter]. Use [▲]/[▼] to select from TONE (CTCSS on transmit only), TSQL((.)) (CTCSS on receive only with beep), TSQL (CTCSS on receive only), DTCS((.)) (DCS with beep), DTCS (DCS), TSQL-R (reverse squelch CTCSS), DTCS-R (reverse squelch DCS), DTCS(T) (DCS on transmit only),

`TONE(T)/DTCS(R)` (CTCSS on transmit, DCS on receive), `DTCS(T)-TSQL(R)` (DCS on transmit, tone on receive) or `OFF` (no tones). Press MENU to exit the menu.

10. Set transmit power: hold V/MHz down and rotate the knob. This will cycle through `SLO` (0.1 W), `L01` (0.5 W), `L02` (1.0 W), `MID` (2.5 W) and nothing displayed (5 W). Release V/MHz to set.

11. Write to a memory: hold M/CALL for one second.

12. Scroll to desired memory using the knob.

13. Hold M/CALL for one second to write.

14. Go to memory mode: press M/CALL (display will show `MR` and a number in lower right).

15. Select the memory you just wrote: use the knob.

Lock/unlock radio

Hold MENU for one second to lock/unlock.

Check repeater input frequency

Hold SQL to listen on the repeater input.

Change power in the field

To set transmit power, hold V/MHz down and rotate the knob. This will cycle through `SLO` (0.1 W), `L01` (0.5 W), `L02` (1.0 W), `MID` (2.5 W) and nothing displayed (5 W). Release V/MHz to set.

Adjust volume

Rotate ring to adjust volume (0–39).

Adjust squelch

Hold SQL and rotate knob to adjust squelch (`OPEN`, `AUTO` and `LEVEL1` through `LEVEL9`). Release SQL to set.

Weird Modes

Radio does not transmit

This radio has a PTT lock. Use MENU, navigate to `Function`, select that and then select `PTT Lock`.

QUICK menu does not have TONE entry

You are in DSTAR mode. Press **FM/DV** to select an FM mode.

Useful Information

Hold **Power** for one second to turn off/on.

This radio has a priority channel. If priority scanning is enabled, the radio will display P in the top line. Press **V/MHz** to turn it off. Use **QUICK** then select PRIO Watch to turn it on.

There are at least two versions of this radio (the ID–31A and the ID–31A PLUS). The PLUS version adds additional DSTAR and repeater search features, but the two radios are programmed similarly.

Factory reset

To reset, push **MENU**. Use **▼** to scroll to the last entry Others. Press **Enter** to select. Use **▲**/**▼** to select Reset, then press **Enter**. Use **▲**/**▼** to choose All Reset (clears everything). Press **Enter**. Press **▲** to select YES, press **Enter**, then **▲**/**▼** to select YES again. Press **Enter** for a full reset.

Settings reset

To reset, push **MENU**. Use **▼** to scroll to the last entry Others. Press **Enter** to select. Use **▲**/**▼** to select Reset, then press **Enter**. Use **▲**/**▼** to choose Partial Reset (clears settings but not memories). Press **Enter**. Press **▲** to select YES, press **Enter** and the radio will reset.

Radio Layout

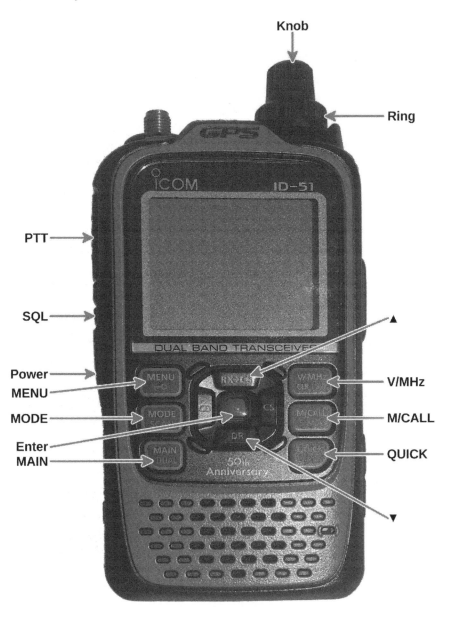

Specs

Receivers Two independent receivers, simultaneous receive
Receives 108–174 MHz FM, DSTAR, AM, 390–479 MHz FM, DSTAR, AM
Transmits 144–148 MHz @ 5 W FM, DSTAR, 430–45- MHz @ 5 W FM, DSTAR
Antenna connector SMA F on radio; needs SMA M antenna
Modes FM, DSTAR
Memory Channels 500
Power 12–16 V DC, 3.5mm OD/1.3mm ID center positive
Model year 2012

Standard Tasks

Program frequency in the field

1. If you aren't already in VFO mode, press [V/MHz]. The screen will show MR or C in the lower right if you aren't.

2. Select the band. Press [QUICK] to open the menu, use [▲]/[▼] to select Band Select, then press [Enter]. Use [▲]/[▼] to choose the band you want, then press [Enter].

3. If needed, set step size: press [QUICK] then use [▲]/[▼] to select TS. Press [Enter]. Use [▲]/[▼] to select the step size you want, then press [Enter].

4. Set mode: press [FM/DV] to cycle through FM (wide FM), FM-N (narrow FM), DV (DSTAR) and AM (no transmit on AM.

5. Set frequency: use the knob. Press [V/MHz] to cycle through 1 MHz adjustments , 10 MHz adjustments or normal step size.

6. Set repeater shift: press [MENU]. Use [▲]/[▼] to navigate to DUP/TONE. Press [Enter]. Use [▲]/[▼] to select DUP. Press [Enter]. Use [▲]/[▼] to select DUP- (negative offset), DUP+ (positive offset) or OFF (simplex). Press [Enter] to set. Press [MENU] to exit the menu..

7. If you need to change the repeater offset frequency: press [QUICK] then use [▲]/[▼] to select Offset Freq. Press [Enter]. Use the knob to adjust offset. Pressing [V/MHz] will cycle through 1 MHz, 10 MHz or step size. Press [MENU] to exit the menu..

8. Set transmit tone: (for transmit), press [MENU]. Use [▲]/[▼] to navigate to DUP/TONE. Press [Enter]. Use [▲]/[▼] to select Repeater Tone. Press [Enter]. Use knob to adjust. Press [MENU] to exit the menu..

9. Set transmit tone: (for receive), press [MENU]. Use [▲]/[▼] to navigate to DUP/TONE. Press [Enter]. Use [▲]/[▼] to select TSQL Freq. Press [Enter]. Use knob to adjust. Press [MENU] to exit the menu.

10. Set transmit tone type: press **QUICK**. Use **▲** **▼** to navigate to TONE. Press **Enter**. Use **▲** **▼** to select from TONE (CTCSS on transmit only), TSQL((.)) (CTCSS on receive only with beep), TSQL (CTCSS on receive only), DTCS((.)) (DCS with beep), DTCS (DCS), DTCS-R (reverse squelch DCS), DTCS(T) (DCS on transmit only), TONE(T)/DTCS(R) (CTCSS on transmit, DCS on receive), DTCS(T)-TSQL(R) (DCS on transmit, tone on receive) or OFF (no tones). Press **MENU** to exit the menu.

11. Set transmit power: hold **V/MHz** down and rotate the knob. This will cycle through SL0 (0.1 W), L01 (0.5 W), L02 (1.0 W), MID (2.5 W) and nothing displayed (5 W). Release **V/MHz** to set.

12. Write to a memory: hold **M/CALL** for one second.

13. Scroll to desired memory using the knob.

14. Hold **M/CALL** for one second to write.

15. Go to memory mode: press **M/CALL** (display will show MR and a number in lower right).

16. Select the memory you just wrote: use the knob.

Lock/unlock radio

Hold **MENU** for one second to lock/unlock.

Check repeater input frequency

Hold **SQL** to listen on the repeater input.

Change power in the field

To set transmit power, hold **V/MHz** down and rotate the knob. This will cycle through SL0 (0.1 W), L01 (0.5 W), L02 (1.0 W), MID (2.5 W) and nothing displayed (5 W). Release **V/MHz** to set.

Adjust volume

Rotate ring to adjust volume (0–39).

Adjust squelch

Hold **SQL** and rotate knob to adjust squelch. Release **SQL** to set.

Weird Modes

Radio does not transmit

This radio has a PTT lock. Use **MENU**, navigate to Function, select that and then select PTT Lock.

Useful Information

Hold [**Power**] for one second to turn off/on.

This radio has dual watch. Press [**MAIN**] to switch which side of the radio will transmit. Hold [**MAIN**] for one second to toggle dual watch on and off.

There are several version of this radio (ID-51, ID-51A, ID-51A Anniversary Edition, ID-51A Plus, ID-51A Plus2) with different colors. All are programmed similarly.

Factory reset

To reset, push [**MENU**]. Use [▼] to scroll to the last entry `Other`. Press [**Enter**] to select. Use [▲]/[▼] to select `Reset`, then press [**Enter**]. Use [▲]/[▼] to choose `All Reset` (clears everything). Press [**Enter**]. Press [▲] to select `YES`, then [▲] one more time to select `YES` again for a full reset.)

Settings reset

To reset, push [**MENU**]. Use [▼] to scroll to the last entry `Other`. Press [**Enter**] to select. Use [▲]/[▼] to select `Reset`, then press [**Enter**]. Use [▲]/[▼] to choose `Partial Reset` (clears settings but not memories). Press [**Enter**]. Press [▲] to select `YES` and the radio will reset.

Radio Layout

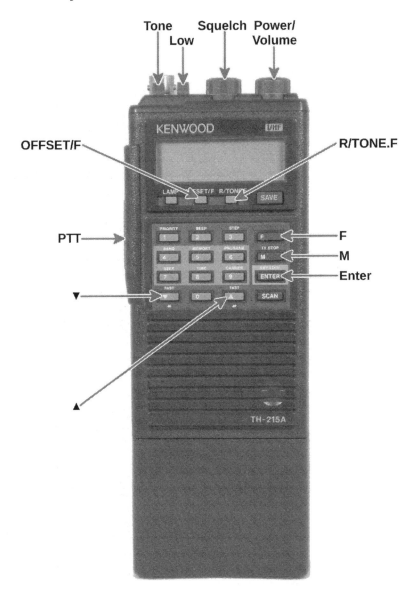

Tone   Squelch Power/
   Low   Volume

OFFSET/F

R/TONE.F

PTT

F

M

Enter

▼

▲

Specs

Receivers Single receiver with priority channel option
Receives 144–147.995 MHz FM
Transmits 144–147.995 MHz FM @ 5 W
Antenna connector BNC F on radio; needs BNC M antenna
Modes FM
Memory Channels 10
Power 7.2–16 V DC barrel style, 5.5 mm OD, 1.55 mm ID plug, center positive
Model year 1987

Standard Tasks

Program frequency in the field

1. Set frequency: press **Enter** then use the number keypad without the first digit (44390 is 144.390 MHz). If needed you can adjust step size by repeatedly pressing **F** then **3** (Step). There is no 12.5 kHz step, so some frequencies may be unavailable.

2. Set repeater shift: press **OFFSET/F**. This will cycle through **+** (plus offset), **–** (minus offset) and no offset.

3. If you need to change the repeater offset frequency: press **F** then **OFFSET/F**. Press **▲** **▼** to choose offset then press **Enter** to set.

4. Set transmit tone type: press **TONE** (button is latching) to enable. To enable tone squelch if TSU-4 board has been installed, rotate the squelch knob all the way counterclockwise (past the click).

5. Set transmit tone: press **F** then **R/TONE.F**. Use **▲** **▼** to choose the frequency, then press **Enter** to set.

6. Set transmit power: press the latching button **LOW**. When the button is up, power is high (5 W on 13.8 V DC, 2.5 W on battery). When the button is down, power is low (0.5 W).

7. Write to a memory: press **M**. The display will show all programmed channels on the left hand side. Press the number of the memory to write (0–9) within five seconds.

8. Select the memory you just wrote: press the memory button (0–9).

Lock/unlock radio

Press **F** then **Enter** to lock/unlock radio.

Check repeater input frequency

Press **R/TONE.F**. Display will show R. Press **R/TONE.F** again to return to normal operation.

Change power in the field

To set transmit power, press the latching button [LOW]. When the button is up, power is high (5 W on 13.8 V DC, 2.5 W on battery). When the button is down, power is low (0.5 W).

Adjust volume

Rotate the power/volume (right) knob to adjust volume.

Adjust squelch

Rotate the squelch (left) knob to adjust squelch. Note that if you rotate the squelch knob all the way counterclockwise (past the click) you will enable tone squelch mode.

Weird Modes

Memory channel indicator blinks

This means the channel has been locked out of the scan. Select that channel, then press [F] [0] to re-enable that channel for scan.

Radio switches to channel 1

Memory 1 can be used as a priority channel. To enable/disable this, press [F] then [1]. Display will show PRIO when priority is enabled.

Useful Information

The Kenwood TH-215A is also available in a 70 cm version, the Kenwood TH-415A and a 1.25 cm version, the Kenwood TH-315A. Programming is similar.

European versions (models ending in E rather than A) do not have a tone encoder. Instead of a [R/TONE.F], they have [REVERSE]. On these radios, hold [REVERSE] to monitor the repeater input. The [TONE] is momentary and sends a 1750 Hz tone.

Tone squelch is available only when an optional TSU-4 board is installed.

The CTCSS tone 97.4 Hz can be used for tone encoding, but not tone squelch.

Factory reset

Hold [F] and [Enter] while turning the radio on. The radio resets immediately.

Kenwood TH-78A

Radio Layout

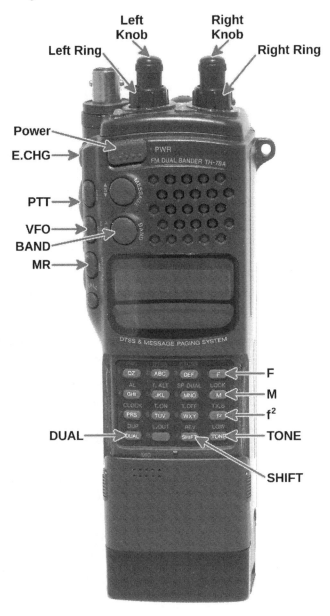

Left Knob
Right Knob
Left Ring
Right Ring
Power
E.CHG
PTT
VFO
BAND
MR
DUAL
F
M
f²
TONE
SHIFT

Specs

Receivers Two independent receivers, simultaneous receive
Receives 118–173.995 MHz FM, 438–449.985 MHz FM
Transmits 144–148 MHz FM @ 5 W, 438–450 MHz FM @ 5 W
Antenna connector SMA F on radio; needs SMA M antenna
Modes FM
Memory Channels 50, 150 with optional ME-1 memory expansion
Power 6.3–16 V DC, Egston 238 barrel style, 3.5mm OD, 1.3mm ID plug, center
     positive
Model year 1992

Standard Tasks

Program frequency in the field

1. If needed, change band: press [BAND].

2. If you aren't already in VFO mode, press [VFO]. The screen will not show a channel number in the bottom center in VFO mode.

3. Set frequency: use the keypad, but omit the first digit (44390 is 144.390 MHz when 2 m band is selected).

4. Set repeater shift: press [SHIFT]. This will cycle through + (plus offset), − (minus offset) and no offset.

5. If you need to change the repeater offset frequency: first turn off the radio. Then hold [SHIFT] while turning the radio on. The display will show SHIFT. Next, hold [F] for at least one second, then release. The display will blink F. Next press [SHIFT]. Use the right knob to adjust, and press [SHIFT] to save. If you see SPLIT instead of SHIFT, turn the radio off and turn it back on while holding down [SHIFT]. Note that there is only one global offset value per band.

6. Set transmit tone type: press [TONE]. The display will show T (send tone on transmit). You can also set tone squelch by pressing [F] then [3/DEF].

7. Set transmit tone: hold [F] for at least one second (F will blink). Press [TONE]. Then use the right knob to adjust. Press [TONE] to set.

8. Set transmit power: repeatedly press [F] then [TONE]. This will cycle through no power level display (high power, 5 W with PB-13 or PB-18 batteries, 2 W with PB-14, PB-17 or alkaline batteries), L (low power, 0.5 W) and EL (economy, 20 mW on 2 m, 10 mW on 70 cm).

9. Write to a memory: press [M]. Enter any two digit memory (01–49) then press [MR] to save.

10. Go to memory mode: press [MR].

11. Select the memory you just wrote: use the right knob.

Lock/unlock radio

Press and release $\boxed{F}$ and then $\boxed{M}$ to lock/unlock radio.

Check repeater input frequency

Press and release $\boxed{F}$ then $\boxed{SHIFT}$ to switch to repeater input frequency (R will be shown in display). Press $\boxed{F}$ then $\boxed{SHIFT}$ again to go back to normal operation.

Change power in the field

To set transmit power, repeatedly press $\boxed{F}$ then $\boxed{TONE}$ This will cycle through no power level display (high power, 5 W with PB-13 or PB-18 batteries, 2 W with PB-14, PB-17 or alkaline batteries), L (low power, 0.5 W) and EL (economy, 20 mW on 2 m, 10 mW on 70 cm).

Adjust volume

Rotate the left knob to adjust volume of the current side (0–20). (Press $\boxed{BAND}$ to switch sides.)

Adjust squelch

Rotate the left ring to adjust squelch of left (2 m) band. Rotate the right ring to adjust squelch of the right (70 cm) band.

Weird Modes

Can't set CTCSS Tone

Most radios sold in North America shipped with the TSU-7 option (tone board) installed. If your radio has no tone board, it will not transmit CTCSS tones.

Most buttons don't work

Assuming the radio is unlocked (no key icon) there is one other reason why keys might not work. This radio has a tone alert function that disables most of the keys. When on, you'll see a bell graphic in the display. Press $\boxed{F}$ and then $\boxed{5/JKL}$ to disable it.

No receive audio

This radio has a DTSS (dual-tone squelch) mode. When enabled, it waits for two DTMF tones before opening squelch. You can turn it off by pressing $\boxed{F}$ and then $\boxed{2/ABC}$.

Knob doesn't tune

The functions of the knobs can be switched. By default, they act as volume and tuning controls for the active band. To switch it, hold $\boxed{F}$ for more than one second (F blinks), then press E.CHG. After this, left and right knobs will act as volume controls for the left and right sides. Repeat the process to switch back.

No buttons

This radio has a sliding panel that protects the buttons from being pushed. If the radio looks like this, slide the panel down by pushing down on both the left and right sides at the same time to expose the buttons.

Panel closed

Useful Information

Press Power to turn the radio on or off.

If you have channel name turned on, you will be able to name channels but will have half the number of memories you usually do. Hold f² while turning the radio on to switch to/from this mode.

Press DUAL to switch between monitoring one frequency and monitoring both frequencies. Use BAND to determine which is the transmit VFO.

To switch to full duplex operation, first switch to dual frequency display. Then hold F for more than one second (F blinks) and then press DUAL. The display will show a blinking DUP. Repeat to turn off.

This radio can operate as a crossband repeater. Hold F for more than one second (F blinks) and then press 0. Repeat the sequence to disable.

Kenwood used a very hard-to-read red for the number buttons. The keypad is:

1/QZ	2/ABC	3/DEF	F/A
4/GHI	5/JKL	6/MNO	M/B
7/PRS	8/TUV	9/WXY	f²/C
#/DUAL	0	*/SHIFT	TONE/D

Factory reset

To reset everything including memories, hold M while turning on the radio.

Settings reset

To reset everything except memories, hold F while turning on the radio.

Radio Layout

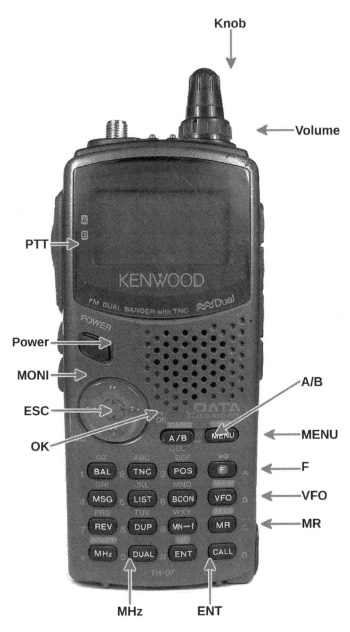

Specs

Receivers Two independent receivers, simultaneous receive
Receives 118–136 MHz AM (band A only), 136–173 MHz FM, 400–479 MHz FM
Transmits 144–148 MHz FM @ 6 W, 430–450 MHz FM @ 5.5 W (using 13.8 V battery)
Antenna connector SMA F on radio; needs SMA M antenna
Modes FM
Memory Channels 200
Power 5.5–16 V DC, Egston 238 barrel style, 3.5mm OD, 1.3mm ID plug, center positive
Model year 1998 for A, 2000 for A(G)

Standard Tasks

Program frequency in the field

1. If needed, change band: press $\boxed{\text{A/B}}$.

2. If you aren't already in VFO mode, press $\boxed{\text{VFO}}$. The screen will show no channel number on the right in VFO mode.

3. Set frequency: press $\boxed{\text{ENT}}$. Enter frequency using the number keypad (144390 is 144.390 MHz).

4. Set repeater shift: press $\boxed{\text{F}}$ then press $\boxed{\text{MHz}}$. This will cycle through + (plus offset), ▬ (minus offset) and no offset. The radio has automatic repeater shift and will usually guess correctly.

5. If you need to change the repeater offset frequency: press $\boxed{\text{F}}$ $\boxed{\text{LIST}}$ (5). Use the knob to adjust. Press $\boxed{\text{OK}}$ to save. Note that this offset will also be used for automatic repeater shift, so if you set a strange offset, put it back to normal after programming the memory.

6. Set transmit tone type: press $\boxed{\text{F}}$ $\boxed{\text{BAL}}$ (1) to turn the tone on (radio shows T) or off.

7. Set transmit tone: press $\boxed{\text{F}}$ $\boxed{\text{TNC}}$ (2). Use the knob to adjust. Press $\boxed{\text{OK}}$ to save.

8. Set transmit power: press $\boxed{\text{F}}$ $\boxed{\text{MENU}}$. This will cycle through H (high, 5 W), L (low, 0.5 W) and EL (economy, 50 mW).

9. Write to a memory: press $\boxed{\text{F}}$ $\boxed{\text{MR}}$. Use the knob to select a memory (empty memories have an empty triangle). Press $\boxed{\text{OK}}$ to save.

10. Go to memory mode: press $\boxed{\text{MR}}$.

11. Select the memory you just wrote: use the knob.

Lock/unlock radio

Hold $\boxed{\text{F}}$ for two seconds to lock/unlock radio.

Check repeater input frequency

Press `REV` (7) to switch to repeater input frequency (R will be shown in display). Press `REV` (7) again to go back to normal operation.

Change power in the field

To set transmit power, press `F` `MENU`. This will cycle through H (high, 5 W), L (low, 0.5 W) and EL (economy, 50 mW).

Adjust volume

Rotate the outer volume ring to adjust volume.

Adjust squelch

Press `F` then press `MONI`. Use knob to adjust squelch. Squelch is separate for each VFO.

Weird Modes

Radio doesn't transmit

This radio has a transmit inhibit function. To disable it, press `MENU` `1` `5` `5` then use the knob to adjust to Off. Then press `OK`.

Radio shows Ch NN

This radio has a channel mode which shows memory numbers and prevents you from entering VFO mode. To disable or enable it, hold `A/B` while turning radio on.

Can't hear one band

Volume for band A and band B can be adjusted so that one is louder than the other. Press `BAL` (1) then use knob to change audio balance (higher is A louder, lower is B louder. At the extreme, the other band will be muted). Press `OK` to set.

Useful Information

Hold `Power` for one second to turn the radio on or off.

This radio also comes in TH-D7A(G) and TH-D7E(G) models. These models have additional APRS functions.

Press `DUAL` (0) to switch between monitoring one frequency and monitoring both frequencies. Use `A/B` to determine which is the transmit VFO.

The ports on the right hand side (from top to bottom) are: earphone (2.5 mm stereo), microphone (3.5 mm stereo), PC cable (2.5 mm stereo), APRS communications (2.5 mm stereo) and 13.8 V DC input (Egston 238 barrel).

Factory reset

To reset the radio to factory defaults, hold F while turning on the radio. Screen will go dark, then show AUX RESET?. Use knob to choose FULL RESET (resets everything). Press OK. Use knob to select YES to do the reset, then press OK to reset, or ESC to abort.

VFO reset

To reset the radio VFOs, hold F while turning on the radio. Screen will go dark, then show AUX RESET?. Use knob to choose VFO RESET (resets VFOs and their settings). Press OK. Use knob to select YES to do the reset, then press OK to reset, or ESC to abort.

Kenwood TH-D72A

Radio Layout

Specs

Receivers Two independent receivers, simultaneous receive
Receives 136–174 MHz FM on VFO A, 410–470 MHz FM on VFO A, 118–174 MHz
    FM on VFO B, 320–524 MHz FM on VFO B
Transmits 144–148 MHz FM @ 5 W, 430–450 MHz FM @ 5 W
Antenna connector SMA F on radio; needs SMA M antenna
Modes FM
Memory Channels 1000
Power 12–16 V DC, Egston 238 barrel style, 3.5mm OD, 1.3mm ID plug, center
    positive
Model year 2010

Standard Tasks

Program frequency in the field

1. If needed, change band: press [A/B].

2. If you aren't already in VFO mode, press [VFO]. The screen will show no memory number on the right in VFO mode.

3. Set frequency: press [ENT]. Enter frequency using the number keypad (144390 is 144.390 MHz).

4. Set repeater shift: press [F] then press [MHz]. This will cycle through + (plus offset), − (minus offset) and no offset. The radio has automatic repeater shift and will usually guess correctly.

5. If you need to change the repeater offset frequency: press [MENU] [1] [6] [0]. Use the knob or [▲]/[▼] to adjust. Press [OK] to save. Press [MENU] to exit the menu mode.

6. Set transmit tone type: press [TONE]. The radio will cycle through T (send tone on transmit), CT (tone on transmit and recieve), DCS (digital coded squelch), D.O (mixed DCS and CTCSS), and off.

7. Set transmit tone: press [F] [TONE]. Use the knob or [▲]/[▼] to adjust. Press [OK] to save.

8. Set transmit power: press [F] [MENU]. This will cycle through H (high, 5 W), L (low, 0.5 W) and EL (economy, 50 mW).

9. Write to a memory: press [F] [MR]. Use the knob or [▲]/[▼] to select a memory (empty memories have an empty triangle). Press [OK] to save.

10. Go to memory mode: press [MR].

11. Select the memory you just wrote: use the knob.

Lock/unlock radio

Hold [F] for two seconds to lock/unlock radio.

Check repeater input frequency

Press [REV] to switch to repeater input frequency (R will be shown in display). Press [REV] again to go back to normal operation.

Change power in the field

To set transmit power, press [F] [MENU]. This will cycle through H (high, 5 W), L (low, 0.5 W) and EL (economy, 50 mW).

Adjust volume

Rotate the outer volume ring to adjust volume.

Adjust squelch

Press [F] then press [MONI]. Use [▲] [▼] or knob to adjust squelch. Squelch is separate for each band.

Weird Modes

Can't set CTCSS Tone/DCS

This is disabled if the radio is not already set up in CTCSS or DCS mode. Use the [TONE] button to make sure you're in the right mode first.

Radio doesn't transmit

This radio has a transmit inhibit function. To disable it, press [MENU] [1] [3] [9] then use the knob to adjust to Off. Then press [OK].

Radio shows Ch NN

This radio has a channel mode which shows memory numbers and prevents you from entering VFO mode. To disable or enable it, hold [PTT] (does not transmit) and [A/B] while turning radio on. Resetting the radio is not possible in channel mode.

Useful Information

Press [Power] for one second to turn the radio on or off.

Press [DUAL] to switch between monitoring one frequency and monitoring both frequencies. Use [A/B] to determine which is the transmit VFO.

The ports on the right hand side (from top to bottom) are: earphone (2.5 mm stereo), microphone (3.5 mm stereo), mini-USB for programming, APRS communications (2.5 mm stereo) and 13.8 V DC input (Egston 238 barrel).

Volume for VFO A and VFO B can be adjusted so that one is louder than the other. Press [MENU] [1] [2] [0] then use [▲] [▼] or knob to change audio balance. Press [OK] to set.

Factory reset

To reset the radio, hold [F] while turning on the radio. Use [▲][▼] or knob to choose `Full Reset` (resets everything). Press [OK] to reset, or [ESC] to abort.

Settings reset

To reset the radio, hold [F] while turning on the radio. Use [▲][▼] or knob to choose `Partial Reset` (resets almost everything except memory channels). Press [OK] to reset, or [ESC] to abort.

VFO reset

To reset the radio, hold [F] while turning on the radio. Use [▲][▼] or knob to choose `VFO Reset` (resets VFOs and their settings). Press [OK] to reset, or [ESC] to abort.

Kenwood TH-D74A

Radio Layout

Specs

Receivers Two independent receivers, simultaneous receive
Receives 136–174 MHz FM/AM/SSB /DSTAR, 216–220 MHz FM/AM/SSB/DSTAR, 410–470 MHz FM/AM/SSB/DSTAR all on band A, 0.1–524 MHz on band B
Transmits 144–148 MHz FM/DSTAR @ 5 W, 222–225 MHz FM/DSTAR @ 5 W, 430–450 MHz FM/DSTAR @ 5 W; transmit on band A only
Antenna connector SMA F on radio; needs SMA M antenna
Modes FM, DSTAR
Memory Channels 1000
Power 11.0–15.9 V DC, Egston 238 barrel style, 3.5mm OD, 1.3mm ID plug, center positive
Model year 2016

Standard Tasks

Program frequency in the field

1. If you aren't already in VFO mode, press **A/B** until you see one frequency only. The screen will show no memory number on the right in VFO mode.

2. Press **MODE** until you see **FM**.

3. Set frequency: press **ENT**, Enter frequency (144390 for 144.390 MHz).

4. Set transmit tone type: press **8** (TONE) to cycle through **T** (send tone on transmit), **CT** (cross-tone, different up and down), **DCS** (digital coded squelch), **D/O** (DCS up, carrier squelch down) and blank (off). You might also see **T/C** (tone up, CTCSS down), **D/C** (DCS up, CTCSS down) or **T/D** (tone up, DCS down).

5. Set transmit tone: press **F** **8** (TONE). Use the knob to adjust the value. Press **A/B** to set.

6. Set repeater shift: press **F** **7** (REV) to cycle through **+** (positive shift), **−** (negative shift) and blank (no shift).

7. If you need to change the repeater offset frequency: press **MENU** **1** **4** **0**. Use the knob to adjust the offset frequency, and press **A/B** to set. Press **MENU** to exit.

8. Set transmit power: press **F** **MENU**. This will cycle through **H** (high, 5 W), **M** (medium, 2 W), **L** (low, 0.5 W) and **EL** (economy, 50 mW).

9. Write to a memory: press **F** **2** (MR) then use the knob or direct keypad entry to select the memory desired. Press **ENT** to write. Press **◄** to go back to the main screen.

10. Go to memory mode: press **2** (MR).

11. Select the memory you just wrote: use the knob.

Lock/unlock radio

Hold [F] for two seconds to lock/unlock radio.

Check repeater input frequency

Press [7] (REV) to switch to repeater input frequency (R will be shown in display).
Press [7] (REV) again to go back to normal operation.

Change power in the field

To set transmit power, press [F] [MENU]. This will cycle through H (high, 5 W), M
(medium, 2 W), L (low, 0.5 W) and EL (economy, 50 mW).

Adjust volume

Rotate the outer volume ring to adjust volume.

Adjust squelch

Press [F] [MONI] and use knob to adjust: Open or 1–5. Press [ENT] to set.

Weird Modes

Can't set CTCSS Tone/DCS

This is disabled if the radio is not already set up in CTCSS or DCS mode. Use the
[TONE] (2) button to make sure you're in the right mode first.

Radio doesn't transmit

This radio has a transmit inhibit function. To disable it, press [MENU] [1] [1] [0]
then use the knob to adjust to Off. Then press [ENT].

Can't hear one side of the radio

Volume for VFO A and VFO B can be adjusted so that one is louder than the other, or
so that one is entirely silent. Press [MENU] [9] [1] [0] then use the knob to change
audio balance. Press [ENT] to set.

Radio display is black or has blinking H

The radio is getting hot or has overheated. Wait for it to cool down before transmitting
again.

Useful Information

Hold [Power] for one second to turn the radio on or off.
    Press [F] [A/B] to switch between monitoring one frequency and monitoring both
frequencies.

This radio has a MicroSD card slot on the right hand side under the speaker mic connector. Under that is a micro USB connector.

Factory reset

To reset the radio, hold [F] while turning on the radio. Use knob to choose Full Reset (resets everything). Press [A/B] to reset.

Settings reset

To reset the radio, hold [F] while turning on the radio. Use knob to choose Partial Reset (resets almost everything except memory channels). Press [A/B] to reset.

VFO reset

To reset the radio, hold [F] while turning on the radio. Use knob to choose VFO Reset (resets VFOs and their settings). Press [A/B] to reset.

Kenwood TH-F6A

Radio Layout

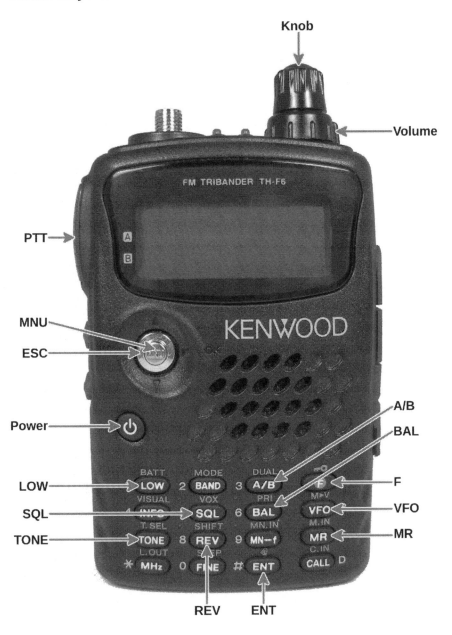

Knob

Volume

PTT

MNU

ESC

Power

A/B

BAL

LOW

F

SQL

VFO

TONE

MR

REV ENT

Specs

Receivers Two independent receivers, simultaneous receive

Receives 136–174 MHz FM on VFO A, 216–260 MHz FM on VFO A, 410–470 MHz FM on VFO A, 0.1–470 MHz AM/FM/SSB on VFO B, 470–824 MHz FM on VFO B, 849–869 MHz FM on VFO B, 894–1300 MHz on VFO B

Transmits 144–148 MHz FM @ 5 W, 222–225 MHz FM @ 5 W, 430–450 MHz FM @ 5 W

Antenna connector SMA F on radio; needs SMA M antenna

Modes AM, FM, SSB

Memory Channels 400

Power 12–16 V DC, Egston 238 barrel style, 3.5mm OD, 1.3mm ID plug, center positive

Model year 2001

Standard Tasks

Program frequency in the field

1. If you aren't already in VFO mode, press [VFO]. The screen will show no memory number on the right in VFO mode.

2. Set frequency: press [ENT]. Enter frequency using the number keypad (144390 is 144.390 MHz).

3. Set repeater shift: repeatedly press [F] then [REV]. This will cycle through + (plus offset), − (minus offset) and no offset. The radio has automatic repeater shift and will usually guess correctly.

4. If you need to change the repeater offset frequency: press [MNU] then use the knob to select menu 6, OFFSET. Press [MNU] then use the knob to change value. Press [MNU] to store. Press [ESC] to exit menu mode.

5. Set transmit tone type: press [TONE]. The radio will cycle through T (send tone on transmit), CT (tone on transmit and recieve), DCS (digital coded squelch), and off.

6. Set transmit tone: press [F] [TONE]. Use the knob to adjust. Press [MNU] to save, then [ESC] to exit menu mode.

7. Set transmit power: press [LOW]. This will cycle through L (low, 0.5 W), EL (economy, 50 mW) and H (high, 5 W).

8. Write to a memory: press [F]. Use the knob to select a memory (empty memories have an empty triangle). Press [MNU] to save.

9. Go to memory mode: press [MR].

10. Select the memory you just wrote: use the knob.

Lock/unlock radio

Hold [F] for two seconds to lock/unlock radio.

Check repeater input frequency

Press [REV] to switch to repeater input frequency (R will be shown in display). Press [REV] again to go back to normal operation.

Change power in the field

To set transmit power, press [LOW]. This will cycle through L (low, 0.5 W), EL (economy, 50 mW) and H (high, 5 W).

Adjust volume

Rotate the outer volume ring to adjust volume.

Adjust squelch

Press [SQL] then use knob to adjust squelch. Squelch is separate for each band.

Weird Modes

Can't set CTCSS Tone/DCS

This is disabled if the radio is not already set up in CTCSS or DCS mode. Use the [TONE] button to make sure you're in the right mode first.

Radio doesn't transmit

This radio has a transmit inhibit function. To disable it, press [MNU] then use the knob to select menu 8, TX INHIBIT. Press [MNU] then use the knob to adjust to OFF. Press [MNU] to save, then press [ESC] to exit.

Radio displays 00:00 or some other time

This radio has "Tone Alert," which displays the time since a signal was received. When it is enabled, many things don't work, and it will beep madly if someone transmits. Press [F] then [ENT] to disable it.

Speaker mic doesn't work

This radio needs to be told what the speaker mic jacks are being used for. Press [MNU] and use the knob to go to menu 9, SP/MIC JACK Press [MNU] and use the knob to select from SP/MIC (speaker mic), TNC (external TNC) and PC (PC programming mode). Press [MNU] to select, then [ESC] to exit.

Radio shows Ch NN

This radio has a channel mode which shows memory numbers and prevents you from entering VFO mode. To disable or enable it, hold [A/B] while turning the radio on.

Don't hear one receiver

Volume for VFO A and VFO B can be adjusted so that one is louder than the other. Press [BAL] then use knob to change audio balance. Press [MNU] to set.

Useful Information

Press [Power] to turn the radio on or off.

Press [F] then [A/B] to switch between monitoring one frequency and monitoring both frequencies. Use [A/B] to determine which is the transmit VFO.

The [MNU] button is a multi-way button. Pushing straight down is [MNU]. Pushing left is [ESC]. Pushing up and down can be used instead of the knob in many cases. Pushing right is [OK].

The TH-F6A and the TH-F7E are roughly identical radios (even sharing the same manual); the TH-F7E was made for Europe and lacks the 1.25 m band.

Factory reset

To reset the radio, hold [F] while turning on the radio. Use the knob to choose FULL RESET (resets everything). Press [MNU] to select, or [ESC] to abort.

Settings reset

To reset the radio, hold [F] while turning on the radio. Use the knob to choose MENU RESET (resets almost everything except memory channels). Press [MNU] to select, or [ESC] to abort.

VFO reset

To reset the radio, hold [F] while turning on the radio. Use the knob to choose VFO RESET (resets VFOs and their settings). Press [MNU] to select, or [ESC] to abort.

Kenwood TH-K20A

Radio Layout

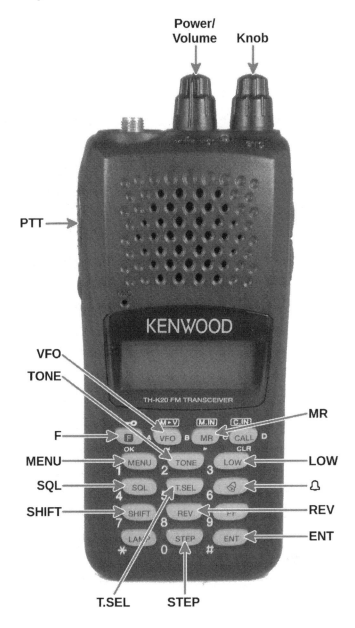

Power/Volume Knob

PTT

VFO

TONE

MR

F

MENU

LOW

SQL

SHIFT

REV

ENT

T.SEL

STEP

Specs

Receivers Single receiver with priority channel option
Receives 136–174 MHz FM
Transmits 144–148 MHz FM @ 5.5 W
Antenna connector SMA F on radio; needs SMA M antenna
Modes FM
Memory Channels 200
Power No DC input on radio, 7.4–9 V DC based on battery options
Model year 2011

Standard Tasks

Program frequency in the field

1. If you aren't already in VFO mode, press **VFO**. The screen will show no memory number on the right in VFO mode.

2. Set frequency: press **ENT** then using the number keypad (144390 is 144.390 MHz). If needed you can adjust step size with **STEP** (0).

3. Set repeater shift: press **SHIFT** (7). This will cycle through **+** (plus offset), **−** (minus offset) and no offset.

4. If you need to change the repeater offset frequency: press **MENU** and use the knob to select menu 7, **OFFSET**. Press **F**. Use the knob to select the offset. Press **F** to set, and then press **MENU** to exit.

5. Set transmit tone type: press **TONE** (2) to select from **T** (CTCSS tone on transmit), **CT** (CTCSS tone on transmit and receive), **DCS** (digital coded squelch) and a triangle symbol (cross tone: separate tones for transmit and receive).

6. Set transmit tone: press **T.SEL** (5). Then use the knob to adjust tone value, and press **T.SEL** (5) to set.

7. Set transmit power: repeatedly press **LOW** (3). This will cycle through no power level display (high power, 5.5 W), **M** (medium power, 2 W), and **L** (low power, 1 W).

8. Write to a memory: press **F** **MR**. Use the knob to select the memory to write, then press **MR** again to write.

9. Go to memory mode: press **MR**.

10. Select the memory you just wrote: use the knob.

Lock/unlock radio

Hold **F** for one second to lock/unlock radio.

Check repeater input frequency

Press [REV] (8). Display will show R. Press [REV] again to return to normal operation.

Change power in the field

To set transmit power, repeatedly press [LOW] (3). This will cycle through no power level display (high power, 5.5 W), M (medium power, 2 W), and L (low power, 1 W).

Adjust volume

Rotate the power/volume (left) knob to adjust volume.

Adjust squelch

Press [SQL] (4) to adjust squelch. Rotate the knob to select from 0–5. Press [SQL] to set.

Weird Modes

Can't enter VFO mode

This radio has a channel-only mode. Hold [PTT] (does not transmit) and [MR] while turning the radio on to exit this mode.

Most buttons don't work

Assuming the radio is unlocked (no key icon) and not in channel mode, there is one other reason why keys might not work. This radio has a tone alert function that disables most of the keys. When on, you'll see △. To disable, press [△]. Use the knob to select OFF; press [△] to set.

Radio doesn't transmit

The Kenwood TH-K20A has a transmit inhibit function. To disable it, press [MENU] and use the knob to select menu 21, TX.INH. Press [F]. Use the knob to select OFF. Press [F] to set, and then press [MENU] to exit.

Useful Information

The Kenwood TH-K20A is also available in a 70 cm version, the Kenwood TH-K40A. Programming is similar.

Factory reset

Hold [F] while turning the radio on. Use the knob to select FL.RST.? (full). Press [F]. You will be prompted with SURE ?. Press [F] to reset. The radio must be unlocked to reset.

Settings reset

Hold [F] while turning the radio on. Use the knob to select PA.RST.? (partial). Press [F]. You will be prompted with SURE ?. Press [F] to reset. The radio must be unlocked to reset.

Radio Layout

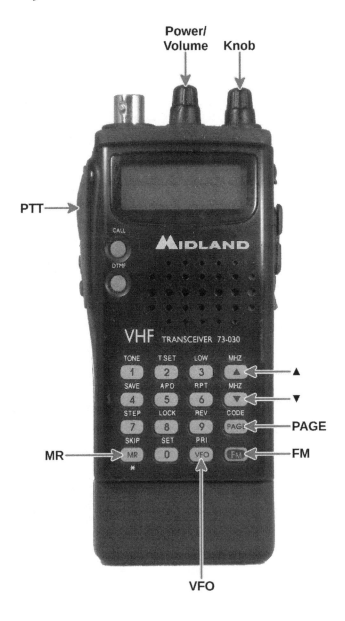

Power/
Volume   Knob

PTT

MR

VFO

▲

▼

PAGE

FM

Specs

Receivers Single receiver with priority channel option
Receives 137–174 MHz FM
Transmits 144–148 MHz @ 5 W FM
Antenna connector BNC F on radio; needs BNC M antenna
Modes FM
Memory Channels 68
Power 5.0–13.8 V DC, EIAJ-02 barrel style, 3.5mm OD, 1.3mm ID plug, **center negative**
Model year 1997

Standard Tasks

Program frequency in the field

1. If you aren't already in VFO mode, press [VFO]. The screen will show **A** in VFO mode, and a memory number in memory mode.

2. Set frequency: use the keypad to enter frequency without leading digit (44390 for 144.390 MHz). You may need to change step size ([FM] then [7] (SET)) to choose your frequency.

3. Set transmit tone type: press [FM] [1] (TONE). Repeatedly pressing [1] (TONE) cycles through **T** (tone on transmit), **CT** (tone on transmit and receive) and blank (no tone).

4. Set transmit tone: press [FM] [2] (T.SET). Use the knob or [▲] [▼] to adjust. Press [2] (T.SET) to set. You will need to work quickly to avoid the timeout. Press [FM] to set.

5. Set repeater shift: press [FM] [6] (RPT). Repeatedly pressing [6] (RPT) cycles through **+** (positive offset), **-** (negative offset) and blank (no shift). Press [FM] to set.

6. If you need to change the repeater offset frequency: press [0] (SET) then use the knob to select **rS:01**. Use [▲] [▼] to adjust. Press [PTT] (does not transmit) to set. You will need to work quickly to avoid the timeout. Note that the repeater offset is global—changing it will change it for all repeater memories.

7. Set transmit power: press [FM] then [3] (LOW). Repeatedly pressing [3] (LOW) cycles through blank (high power 5 W), **L** (low power, 0.6 W) and extra low power **EL**, 0.18 W. Press and release [FM] to set power level. Note that the power level is global—changing it will change it for all memories.

8. Write to a memory: hold [FM] for at least half a second.

9. Use the knob to select the desired memory within five seconds.

10. Press and release [FM] to write.

11. Go to memory mode: press [MR].

12. Select the memory you just wrote: use the knob.

Lock/unlock radio

Press and release [FM] then [8] (LOCK) to lock/unlock.

Check repeater input frequency

Press and release [FM] then [9] (REV) to switch between reverse and regular modes. In reverse mode, display shows R.

Change power in the field

To set transmit power, press [FM] then [3] (LOW). Repeatedly pressing [3] (LOW) cycles through blank (high power, 5 W), L (low power, 0.6 W) and extra low power EL, 0.18 W. Press and release [FM] to set power level. Note that the power level is global—changing it will change it for all memories.

Adjust volume

Rotate the center power/volume knob to adjust volume.

Adjust squelch

Press [FM] then [MONI]. Rotate the knob to adjust squelch (0–9). Press [FM] to save.

Weird Modes

Can't enter VFO mode

This radio has a memory-only mode. To enable/disable it, turn the radio off. Hold [FM] and [PAGE] while turning the radio on. Release the buttons.

Radio shows DT, PAG or T.PAG

This radio has a DTMF paging capability. Repeatedly press [PAGE] to disable.

Radio blinks P

You are monitoring the priority memory (memory 0). Turn it off/on by pressing [FM] then [VFO].

Useful Information

You must have the RTN102A tone squelch unit installed for tone squelch CT to work.

Options in the set menu [FM] [0] (SET) include:

`rS 01` Repeater offset (`060` for 600 kHz)	`bP 09` Enable keypad beep
`AL 02` Number of rings when paging	`bn 10` Keypad beep plays beeps (`be`) or DTMF (`dt`)
`to 03` Time out timer (seconds)	`AS 11` Automatic repeater shift enable
`Pt 04` Delay before transmitting DTMF paging tones (ms)	`En 12`: Prevent changing frequency with knob/changing squelch when radio is locked
`dS 05` DTMF tone playback speed (`F` for 50 ms tones, `S` for 80 ms tones)	`bL 13` Busy channel lock out
`dn 06` DTMF playback mode (`dt` for tones, `be` for beeps)	`tS 14` Prevent transmit (`on`) or allow transmit (`oFF`)
`tH 07` Keep transmitting when entering DTMF even if PTT isn't pressed	`Ab 15` Transmit automatically when DTMF tone received
`SC 08` Scan resume (`to` resumes scanning after five seconds, `Co` resumes scanning when carrier drops)	`dd 16` Show DTMF tone received instead of first digit of frequency

The radio manual refers to this radio as the RL-105, and indicates there may be a UHF version. This radio may also be known as the Maha/Rexon RL-115 and the Midland/Alan CT-22.

This radio has the option to enable transmit and receive on wider frequencies than just the amateur bands. To change the transmit and/or receive range of the radio, press PAGE. Press ▲ / ▼ to select range. Turn the radio off. Press and hold ▲ and ▼ simultaneously while turning the radio on. This will reset the radio's configuration, and change the available frequencies for tuning. Options are:

PAGE	Transmit (MHz)	Receive (MHz)
C	144–146	144–146
1	144–148	144–148
2	144–146	136–174
3	136–174	136–174
4	100–175	100–175
5	138–175	138–175
6	145–175	145–175
P	140–160	140–160

Factory reset

To reset everything, turn the radio off. Hold VFO and ▼ while turing the radio on. Release the buttons.

Settings reset

To reset some settings but leave memories intact, turn the radio off. Hold VFO and ▲ while turning the radio on. Release the buttons.

Radio Layout

Power/
Volume

PTT

BAND

MON

CALL

A/B

VFO/MR

MENU

▲

▼

EXIT

*

#

Specs

Receivers Single receiver, dual watch (first to break squelch wins)
Receives 136–174 MHz and 400–480 MHz or 400–520 MHz FM
Transmits 136–174 MHz @ 4 W FM and 400–480 or 400–520 MHz @ 4 W FM
Antenna connector SMA **M** on radio; needs SMA **F** antenna
Modes FM
Memory Channels 128
Power No DC input on radio
Model year 2017

Standard Tasks

Program frequency in the field

1. This radio doesn't allow you to overwrite memories. You will need to delete first if you want to write to a memory that has data in it. To delete: press `MENU` `2` `8` `MENU` (DEL-CH) *XXX* where *XXX* is the channel (000–127). Then press `MENU` `EXIT`.

2. If you aren't already in VFO mode, press `VFO/MR`. The screen will not display channel numbers on the right hand side in VFO mode.

3. Set frequency: use the keypad (144390 for 144.390 MHz). You may need to change step (menu 1) to enter the value you want.

4. Set transmit tone: press `MENU` `1` `3` `MENU` (T-CTCS) and use `▲` `▼` to select the correct CTCSS tone frequency (or OFF). Then press `MENU` `EXIT`. Menu 12 will let you set DCS instead.

5. Set repeater shift: press `MENU` `2` `5` `MENU` (SFT-D) and use `▲` `▼` to select the correct repeater shift (+, - or OFF). Press `MENU` `EXIT`.

6. If you need to change the repeater offset frequency: press `MENU` `2` `6` `MENU` (OFFSET) and enter the offset using the keypad (00600 for 600 kHz, 05000 for 5 MHz). Then press `MENU` `EXIT`.

7. Set transmit power: press `MENU` `2` `MENU` (TXP) and use `▲` `▼` to select the desired power level. HIGH is 4 W, LOW is 1 W. Then press `MENU` `EXIT`.

8. Write to a memory: press `MENU` `2` `7` `MENU` (MEM-CH) and enter channel to write using the keypad *XXX* (000–127). Then press `MENU` `EXIT`.

9. Go to memory mode: press `VFO/MR`.

10. Select the memory you just wrote: use `▲` `▼`.

Lock/unlock radio

Hold `#` for two seconds to lock/unlock.

Check repeater input frequency

Press [*] to switch into or out of reverse mode (display shows R).

Change power in the field

To set transmit power, press [MENU] [2] [MENU] (TXP) and use [▲][▼] to select the desired power level. HIGH is 4 W, LOW is 1 W. Then press [MENU] [EXIT].

Adjust volume

Rotate power/volume knob to adjust volume.

Adjust squelch

Press [MENU] [0] [MENU] (SQL) then use [▲][▼] to select the correct squelch level (0 is open, 1–9 are available levels), then press [MENU] [EXIT].

Weird Modes

Can't leave channel (memory) mode

Some distributors ship these radios with VFO mode turned off. If that's the case, you will need to modify the programming with a computer/radio programming cable to turn VFO mode on. This cannot be changed from the front panel.

Can't set offset/direction

This radio can display frequencies instead of channel names. If you have that enabled, it's difficult to tell if you're in channel (memory) mode or frequency (VFO) mode. One sure way is to try to program an offset or shift. If it doesn't "stick" after programming, that's likely because you're in channel mode. Switch to frequency mode in order to program.

Can't reset radio

The radio can be set so that you can't perform a reset. If that's the case, you will see menu 40 but not be able to select it with [MENU]. This can be changed from the SET menu by entering -DD-DDD using [VFO/MR]. (Note: This corrupts the ANI ID.)

Can't leave VFO mode

If you have no memories programmed, you will not see channel numbers when you try to enter channel (memory) mode. If you have menu menu 14, VOICE set to ENG you will hear "Frequency Mode" or "Cancel" rather than "Channel Mode" when you push [VFO/MR]. The easiest way around this is to program a channel.

SET mode

Hold [1] while turning the radio on to enter a SET menu. This appears to let you specify the freqencies that the radio will accept. When the display shows SET, press [MENU] and you'll see the VHF range (normally 136–174). To change it, enter a new range using the number buttons. Press [MENU] to go to the next entry. The dash in the middle is not counted as a digit. (Pressing [1][MENU][3][MENU][6][MENU][1] [MENU][7][MENU][4][MENU] will reset it to default.) Then do the same with the UHF range (which by default is 400–520). The radio will reset, and your range will be changed. If you don't wan't to change the values, press [MENU] six times.

It appears that features can be enabled with SET mode as well. Press [MENU][MENU] and then [VFO/MR] will show D, [A/B] will show E, and [BAND] will show H, [*] will show ;, [#] will show >, [MON] will show G, [CALL] will show F.

Setting with the SET menu appears to corrupt the ANI ID.

Useful Information

If you wait too long after pressing [MENU], the radio will time out. You need to act quickly when programming.

This radio comes in multiple colors.

Factory reset

Press [MENU][4][0][MENU] (RESET) then use [▲][▼] to select ALL and press [MENU]. The display will show SURE?. Press [MENU] to reset. The reset for this device is not perfect, and may fail to reset all settings.

VFO reset

Press [MENU][4][0][MENU] (RESET) then use [▲][▼] to select VFO and press [MENU]. The display will show SURE?. Press [MENU] to reset. The reset for this device is not perfect, and may fail to reset the VFO.

⚠ WARNING: In at least some configurations, the Quansheng UV-R50 may permit you to transmit on business or public safety frequencies. Make sure you are in-band when transmitting.

Radio Layout

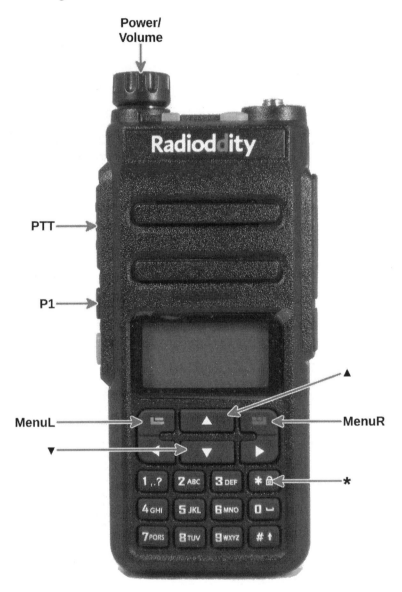

Specs

Receivers Single receiver, dual watch (first to break squelch wins)
Receives 136–174 MHz FM/DMR, 400–470 MHz FM/DMR
Transmits 136–174 MHz FM/DMR @ 5 W, 400–470 MHz FM/DMR @ 5 W
Antenna connector SMA F on radio; needs SMA M antenna
Modes FM, DMR
Memory Channels 1024
Power No DC input on radio; 12 V DC, 5.5 mm OD, 2.0 mm ID, center positive on
    charger base
Model year 2017

Standard Tasks

Program frequency in the field

This radio does not allow you to add new frequencies in the field. However, you can modify existing frequencies that have been programmed. Many codeplugs have an "FPP" zone intended for this purpose.

⚠ **WARNING:** This radio can be configured to prevent front panel programming, or to allow it only with a password. If it's configured this way, you will not be able to program the radio.

1. On this radio, you select the destination memory first. To select: use `▲`/`▼`. Note that you can't convert an analog FM memory into a DMR memory or vice versa. You may need to change zones; see "Useful Info" below.

2. To enter programming mode, press `MenuL` `▲`/`▼` `MenuL` to select the `Set` menu. Then press `▲`/`▼` to select menu 3, `Radio Cfg`. Press `MenuL` to select.

3. Set receive frequency: use `▲`/`▼` to select `Rx_Freq` and press `MenuL`. Press `MenuR` repeatedly to clear out the frequency, and enter your new frequency, using five digits after the decimal point (144.390 MHz is 14439000). Press `MenuL` to set.

4. Set transmit frequency: use `▲`/`▼` to select `Tx_Freq` and press `MenuL`. Press `MenuR` repeatedly to clear out the frequency, and enter your new frequency, using five digits after the decimal point (144.390 MHz is 14439000). Press `MenuL` to set. Press `MenuR` repeatedly to exit the menu system.

5. Set receive tone type and value: Press `MenuL` `▲`/`▼` `MenuL` to select the `Set` menu. Then press `▲`/`▼` to select menu 1, `Radio Set`. Press `MenuL` to select. Press `▲`/`▼` to choose `CTCSS/DCS` then press `MenuL`. Press `▼` to choose menu 2, `R CTC/DCS` and press `MenuL`. Use `▲`/`▼` to select the value or `Off`. Press `MenuL` to set. Press `MenuR` repeatedly to exit the menu system.

6. Set transmit tone type and value: press `MenuL` `▲`/`▼` `MenuL` to select the `Set` menu. Then press `▲`/`▼` to select menu 1, `Radio Set`. Press `MenuL` to select. Press `▲`/`▼` to choose `CTCSS/DCS` then press `MenuL`. Press `▼` to

choose menu 3, **T CTC/DCS** and press **MenuL**. Use **▲**/**▼** to select the value or **Off**. Press **MenuL** to set. Press **MenuR** repeatedly to exit the menu system.

7. Set transmit power: press **MenuL** **▲**/**▼** **MenuL** to select the **Set** menu. Then press **▲**/**▼** to select menu 1, **Radio Set**. Press **MenuL** to select. Press **▲**/**▼** to choose **TX Power** then press **MenuL**. Press **▲**/**▼** to select **Low** (1 W) or **High** (5 W). Press **MenuL** to set. Press **MenuR** repeatedly to exit the menu system.

Lock/unlock radio

Hold ***** for two seconds to lock. Press **MenuL** ***** to unlock when locked. (Note that the radio can be configured to auto-lock by going into the **Radio Set** menu with **KeyPad Lock**.)

Check repeater input frequency

This radio has no option to listen to the repeater input frequency. You will have to program a separate memory with the repeater input frequency to do this.

Change power in the field

To set transmit power, press **MenuL** **▲**/**▼** **MenuL** to select the **Set** menu. Then press **▲**/**▼** to select menu 1, **Radio Set**. Press **MenuL** to select. Press **▲**/**▼** to choose **TX Power** then press **MenuL**. Press **▲**/**▼** to select **Low** (1 W) or **High** (5 W). Press **MenuL** to set. Press **MenuR** repeatedly to exit the menu system.

Adjust volume

Rotate the the power/volume knob to adjust the volume.

Adjust squelch

Press **MenuL** then **▲**/**▼** then **MenuL** to select **Radio Set**. Then press **MenuL** to choose menu 4, **Talkaround**. Press **MenuL** to select. Use **▲**/**▼** to select the squelch you want (**Strict** or **Normal**). Press **MenuL** to set.

Weird Modes

Can't set tone while programming

This radio doesn't allow you to program an analog FM frequency into a DMR memory or vice versa. You may need to change zones; see "Useful Info" below.

Menus don't match instructions

The GD-77 has an open source firmware version "OpenGD77" which is programmed differently from the stock firmware. If you have a **Credits** menu item above the **Zone** menu item, you are using the open source firmware. The stock CPS will not program open source firmware.

Useful Information

The radio's memories are divided into zones (similar to banks), with a maximum of sixteen memories per bank. To change zones, press MenuL then ▼ until you select Zone. Press MenuL to confirm. Press ▲ ▼ to select a zone, then press MenuL to confirm. To change memories within a zone, use ▲ ▼.

The radio menu system has a ten-second timeout. You will need to act quickly once you have entered the menu.

Factory reset

To reset the radio, hold 1 and P1 while turning the radio on. The radio will display Memory Reset? Press MenuL to reset. After a factory reset, channel 16 in zone 1 and channels 15 and 16 in zone 2 are analog frequencies.

⚠ WARNING: In at least some configurations, the Radioddity GD-77 may permit you to transmit on business or public safety frequencies. Make sure you are in-band when transmitting.

Radio Layout

Power/
Volume

Squelch

MO

REV

PTT

VHF FM

RadioShack

LOCK

FUNC

SC

MR

BEEP

168

Specs

Receivers Single receiver
Receives 136–174 MHz FM
Transmits 142–149.885 MHz @ 200 mW FM
Antenna connector SMA F on radio; needs SMA M antenna
Modes FM
Memory Channels 30
Power 9 V DC, "I" barrel style jack 3.8 mm OD, 1.1 mm ID center positive
Model year 1998

Standard Tasks

Program frequency in the field

1. On this radio, you select the destination memory first. To select: press [MR] if you aren't already in memory mode. The screen will show MR in the upper right when you are in memory mode. Press [▲][▼] to select the memory you want to write.

2. Hold [FUNC] and press [MR]. The memory will flash.

3. Set frequency: hold [FUNC] and press [▲]. Press [▲][▼] to change digit. Hold [FUNC] and press [▲] to move to next digit.

4. If you need to change the repeater offset frequency: hold [FUNC] and press [SC]. The screen will show rPt. Use [▲][▼] to select offset in MHz. (600 kHz is 0.6.) Use an offset of 0.0 for simplex.

5. Set repeater shift: hold [FUNC] and press [MO]. Repeating will cycle through +, − and none—displayed above the second digit in the frequency.

6. Set transmit tone type and value: hold [FUNC] and press [▼]. The display will show tONE oF. Press [▼] until display shows tONE oN. Next hold [FUNC] and press [▼]; you will see rC (receive tone). Adjust it with [▲][▼]; oFF means no tone. Then hold [FUNC] and press [▼] to see tC (transmit tone). Adjust the value using [▲][▼].

7. Write to a memory: press [PTT]. Note that if you leave memory mode (stops flashing) you will have to re-enter it by holding down [FUNC] and pressing [MR]. If you hold [FUNC] down for too long without pressing another key, the memory will be deleted.

Lock/unlock radio

Hold [FUNC] and press [C] (LOCK) to lock/unlock. This locks everything except PTT, volume, squelch and backlight. Key symbol will be displayed when radio is locked.

Check repeater input frequency

Hold FUNC and press PTT (does not transmit). Display will change to the input frequency. There is no other indication that you're in reverse, so be cautious. Repeat procedure to switch back.

Change power in the field

There is no setting for power level. Power is 200 mW on batteries, or 2 W on 9 V DC.

Adjust volume

Rotate inner power/volume knob to adjust volume.

Adjust squelch

Rotate outer squelch ring to adjust squelch.

Weird Modes

Display shows all LCD segments or PrESS

This radio has a diagnostic mode that you can enter by pushing MO while turning on the radio. If the display is showing all LCD segments or PrESS followed by a few letters, turn the radio off and then on again to skip the diagnostic.

Display shows EEP-Error

The EEPROM info is bad. You might be able to reset the radio using the reset procedure described below.

Display shows PLL-Error

The PLL couldn't get a lock. Power off and power on again.

Display shows S-SHORT

The external microphone has a short. Remove the external mic and use the internal mic.

Display shows Inhibit on PTT

You are trying to transmit outside of the range allowed by the radio. Check your offset.

Radio doesn't transmit

The radio has a busy channel lock-out mode which prevents transmission when there is a carrier on the frequency. If it has been enabled, you can disable it with the following procedure. Turn the power on while holding FUNC. Hold FUNC and press ▼ until you see bCLO-oN. Press ▼ to turn it off. Press PTT (does not transmit) to set.

Useful Information

Holding down [MO] will open squelch for as long as you hold it down.

You can see the current receive tone (rc), transmit tone (tc), memory lockout (SCSP), repeater offset MHz (rpt) and frequency step MHz (CS) by holding the [MO] button down for more than one second.

By default, the radio transmits from 144.0 through 148.0 MHz. To expand the transmit range to 142.0 through 149.885, hold [SC] when turning on the radio. The range will be shown briefly in the display. Repeating the operation restricts the transmit range again.

The configuration menu you get when you hold [FUNC] and turn on the power has the following options: CS (frequency step), bCLO (busy channel lock-out), t_dY (drop CTCSS tone before ending transmission to prevent squelch tail), Sd (scan delay in seconds), tot (time-out timer in seconds), PS (power save), rPt (VFO repeater shift). Hold [FUNC] and press [▼] to cycle through them. Press [PTT] (does not transmit) to exit.

Programming the VFO for repeater operation is a little tricky (need to set VFO offset in config menu) and not necessary. Just write a memory, and use the VFO for simplex only.

Factory reset

To reset the radio, hold [FUNC] and [MO] while turning the radio on. The display will show InitiAL as it resets. This also clears all memories.

Radio Layout

(Image is of 1994 model)

Specs

Receivers Single receiver
Receives 144–148 MHz FM
Transmits 144–148 MHz @ 6 W FM
Antenna connector BNC F on radio; needs BNC M antenna
Modes FM
Memory Channels 12 standard
Power 7.2–13.8 V DC per spec, "K" barrel style jack 5 mm OD, 2.1 mm ID center
  positive
Model year 1992 (Realistic markings), 1994 (Radio Shack markings)

Standard Tasks

Program frequency in the field

1. If you aren't already in VFO mode, press [**D**](VF). The screen will show **M-CH** if you aren't in VFO mode.

2. Set frequency: enter the last four digits of on the numeric keypad (so 144.390 MHz is 4390).

3. Set repeater shift: hold [**F**] and press [**3**](+/–) until correct direction appears. (Note that this radio gets repeater shift wrong for 147 MHz, so be careful.).

4. If you need to change the repeater offset frequency: hold [**F**] and press [**8**](M-SET). Display will show **oS**. Use the right knob to adjust offset; default is **0.600**. Press [**PTT**](does not transmit) to set.

5. Set transmit tone type: hold [**F**] and press [**1**](T-SQL). Display will show **T-SQL**.

6. Set transmit tone: hold [**F**] and press [**8**](M-SET). Radio will show **tF** (transmit frequency). Press [▼] to see **tc** (transmit tone). Rotate the right knob to set the value or use **oFF** for no transmit tone. Press [▼] to see **rc** (receive squelch tone). Rotate the right knob to set the value or use **oFF** for no receive squelch tone. Press [**PTT**](does not transmit) to set.

7. Write to a memory: hold [**F**] and turn the right knob until the memory number you want to write to is displayed. Release [**F**]. Hold [**F**] and then press [**C**] (M-WR) for one second to write.

8. Go to memory mode: press [**C**](MR).

9. Set transmit power: press the latching button [**Low Power**]. Button down is low power (about 1 W); button up is high power (between 2.5 W and 6 W depending on power source).

Lock/unlock radio

Hold [**F**] and press [**A**](LOCK) to lock/unlock. The lock disables everything except PTT, Volume and Squelch.

Check repeater input frequency

Hold [F] and press [6] (REV). Display will switch from receive frequency to transmit frequency, and repeater shift will change from **+** to **–** or vice versa. Repeat procedure to switch back.

Change power in the field

To set transmit power, press the latching button [**Low Power**]. Button down is low power (about 1 W); button up is high power (between 2.5 W and 6 W depending on power source).

Adjust volume

Rotate the center power/volume knob to adjust volume.

Adjust squelch

Rotate the left Squelch knob.

Weird Modes

Display shows ER1

This means the lithium battery that maintains memory when the radio is powered off has died. Use the reset procedure described below to bypass the error condition.

Display shows ER2

This means the PLL is unable to lock. The radio will need repair.

Radio doesn't transmit

The HTX-202 can be set to prevent transmission. If it's set, the lower left corner of the LCD will display INH. To turn transmissions back on, make sure you're in VFO mode by pressing [D] (VF). Then hold [F] and press [8] (M-SET). Press [▼] eleven times until you see tE. Use the right knob to adjust between oFF (no transmission) and on (transmission allowed). Press [PTT] (does not transmit) to set.

Radio doesn't receive

The HTX-202 can be set to require a DTMF sequence before it opens the squelch. If this is turned on, DTMF will appear in the lower right of the display. To turn it off, hold [F] and press [4] (D-SQL).

Useful Information

You can see the transmit frequency, tone and tone squelch values by holding the [M] button (above [PTT] and below [F]).

In VFO mode, the menu you get when you hold [F] and press [8] M–SET has the following options: oS (offset), tc (default transmit tone value), rc (default receive tone squelch value), Sr (frequency step), Sc (scan resume), Sd (scan delay), S1 (lower scan limit), S2 (upper scan limit), ud (scan direction), PS (power save—higher number is less power used), tE (transmit enable), to (transmit timeout), Lb (priority channel scan time), Ar (touch tone auto reply). In memory mode, you'll see tF (transmit frequency for current memory), tc (transmit tone value for current memory) and rc (receive tone squelch for current frequency).

⚠ **WARNING:** The belt clip of the HTX-202 also doubles as the heat sink for the final transistors. Don't operate without the belt clip in place.

⚠ **WARNING:** Always remove the battery pack before operating on external power.

Factory reset

To reset the radio (and clear an ER1 condition), hold [D] (CLR) and [F] while turning the radio on. This also clears all memories and settings. The 1992 model with Realistic markings resets to 146.200; the 1994 model with Radio Shack markings resets to 146.600.

Radio Shack HTX-245

Radio Layout

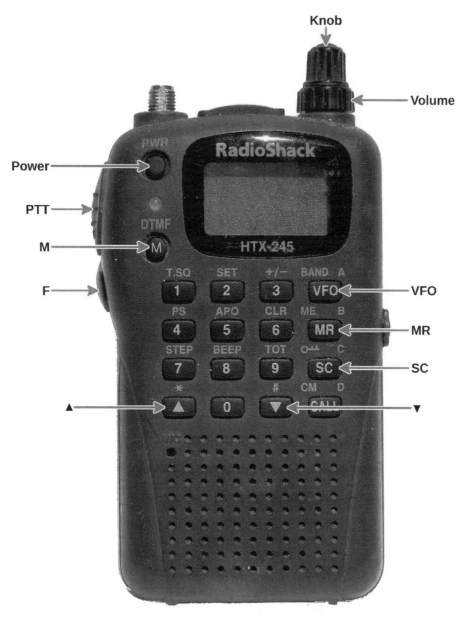

Specs

Receivers Single receiver
Receives 144–148 MHz FM, 438–450 MHz FM
Transmits 144–148 MHz @ 1.5 W FM, 438–450 MHz @ 1.5 W FM
Antenna connector SMA F on radio; needs SMA M antenna
Modes FM
Memory Channels 50
Power 4.5–6 V DC, "H" barrel style jack 3.4 mm OD, 1.3 mm ID center positive
Model year 2000

Standard Tasks

Program frequency in the field

1. If you aren't already in VFO mode, press **VFO**. The screen will show **MR** if you aren't in VFO mode.

2. If needed, change band: hold **F** and press **VFO** to scroll through 2 m, NOAA weather and 70 cm.

3. Set frequency: use the numeric keypad (144.390 MHz is entered as 144390).

4. Set repeater shift: hold **F** and press **3** (+/−) until correct direction appears (**+** is positive, **−** is negative, blank is simplex).

5. Set transmit tone type: hold **F** and press **1** (T-SQL) to cycle through **T** (tone), **TSQ** (tone squelch) and blank (no tone).

6. Set transmit tone: hold **F** and press **2** (SET). Display will show **Sq** (looks like 59 to most of us). Use knob to scroll to **rt** (receive tone) and/or **tt** (transmit tone). Use **▲**/**▼** to adjust.

7. If you need to change the repeater offset frequency: note that there are two offsets in the radio. One defaults to 600 kHz; the other to 5 MHz. While in set mode use the knob to scroll to **r1** (2 m offset, default 0.60) or **r2** (70 cm offset, default 5.00). Use **▲**/**▼** to adjust.

8. Hold **F** and press **2** (SET) to exit the menu system.

9. Write to a memory: hold **F** then press and release **MR**. Turn knob until the memory number you want to write is displayed. Hold **F** and press **MR** to write.

10. Go to memory mode: press **MR**. You should be on the channel you just wrote.

11. Select the memory you just wrote: use the knob.

Lock/unlock radio

Hold **F** and press **SC** to lock/unlock. The lock disables everything except PTT, volume, monitor and power.

Check repeater input frequency

This radio has no option to listen to the repeater input frequency. You will have to program a separate memory with the repeater input frequency to do this.

Change power in the field

This radio has only one power level. On 6 V DC external supply it provides 1.5 W output. On 4.5 V DC internal batteries it provides 700 mW.

Adjust volume

Rotate the ring to adjust volume.

Adjust squelch

Press [VFO], then hold [F] and press [2] (SET) to enter Set mode. Rotate knob until you see Sq (looks like 59 to most of us). Press [▲]/[▼] to set it from 0–5. Hold [F] and press [2] (SET) to exit.

Weird Modes

Can't change parameters

Some parameters require you to be in VFO mode to change them. Press [VFO] before pressing the buttons you want.

Useful Information

Hold [Power] for one second to turn the radio on or off.

As well as monitoring, the [M] button will display the transmit frequency, tone and tone squelch values when you hold it down.

⚠ **WARNING:** There is a switch in the battery compartment which controls whether or not an external adapter charges the batteries. For NiCd or NiMH, the power switch should be set to "On." For other batteries, the power switch should be set to "Off."

This radio has two versions, part numbers 19-1106 and 19-1106A. The "A" version can be given extended transmit/receive frequency ranges (142.00–149.88 MHz and 420–450 MHz). To do this, hold [2] and [3] and turn the radio on. The non-"A" version requires a hardware mod.

Factory reset

To reset the radio hold [6] (CLR) and [MR] while turning the radio on. This clears all memories and settings.

Radio Layout

Specs

Receivers Single receiver with priority channel option
Receives 108–136.9875 MHz AM, 137–174 MHz FM, 420–512 MHz FM
Transmits 144–148 MHz @ 4 W FM, 438–450 MHz @ 4 W FM
Antenna connector SMA F on radio; needs SMA M antenna
Modes FM
Memory Channels 100
Power 5.0–13.8 V DC, "H" barrel style jack 3.4 mm OD, 1.3 mm ID center positive
Model year 2002

Standard Tasks

Program frequency in the field

1. If you aren't already in VFO mode, press [MR]. The screen will show MR if you aren't.

2. If needed, change band: press [Enter] repeatedly to go through 2 m, 70 cm, air band and NOAA weather.

3. If needed, set step size: hold [F] and press [2] (STEP) to adjust step if necessary. Use [▲][▼] to select step, and [Enter] to set.

4. Set frequency: use the numeric keypad (144.390 MHz is entered as 144390).

5. Set repeater shift: hold [F] and press [9] (+/−) until correct direction appears (**+** is positive, **−** is negative, blank is simplex).

6. Set transmit tone: press [▶] until you see **TXTONE** (or **RXTONE** for receive). Then press [▲][▼] to adjust, then [Enter] to set.

7. Set transmit tone type: hold [F] and press [1] (T.SQL). This will cycle through **T** (transmit tone), **SQ** (receive tone), **TSQ** (both send and receive tone) and blank (no tone).

8. If you need to change the repeater offset frequency: note that there are two offsets in the radio, one for VHF frequencies and another for UHF frequencies. To change these values, press [▶] until you see **VHFRPT** or **UHFRPT**. Then press [▲][▼] to adjust, and [Enter] to set.

9. Set transmit power: press [H/L]. This will cycle through **L** (1 W), **M** (approximately 1.5 W on batteries, 2 W on external power) and blank (high power: 3 W on batteries, 4 W on external power).

10. Write to a memory: hold [F] then press and release [MR]. Use [▲][▼] or the knob until the memory number you want to write is displayed. Hold [F] and press [MR] to write.

11. Go to memory mode: press [MR]. You should be on the channel you just wrote.

12. Select the memory you just wrote: use the knob.

Lock/unlock radio

Hold [F] and press [DW] to lock/unlock. The lock disables everything except PTT, volume, monitor and power.

Check repeater input frequency

Hold [F] and press [8] (REV) to switch to reverse mode (display shows opposite offset, but otherwise gives no indication you're in reverse). Hold [F] and press [8] (REV) to return to normal.

Change power in the field

To set transmit power, press [H/L]. This will cycle through L (1 W), M (approximately 1.5 W on batteries, 2 W on external power) and blank (high power, 3 W on batteries, 4 W on external power).

Adjust volume

Rotate the ring to adjust volume.

Adjust squelch

Press [▶]. Display will show SQL. Press [▲][▼] to adjust, then press [Enter] to set.

Weird Modes

Display shows ERR

The radio will show ERR when it loses PLL lock. This can happen if RF gets into the PLL circuit. Turn off the radio and turn it back on to clear it; if it happens regularly reduce power.

If the radio shows ERR at other times, it may need service.

Useful Information

Press [Power] to turn the radio on or off.

The external power socket on the radio just powers the radio. The battery has its own power socket (same style) that charges the battery. The charging voltage is 12.0 V DC.

The radio can transmit on one frequency and monitor another for satellite operation. Press [MR] to go to VFO mode, then enter your receive frequency. Then Hold [F] and press [SC] (display shows XB). Next, enter your transmit frequency. You can hold [F] and press [8] to switch frequencies. Hold [F] and press [SC] to leave this mode.

This radio comes with a compass. To use it, hold [F] and press [H/L] (COMP). The radio must be roughly horizontal to use the compass.

To calibrate the compass, place the radio flat on its back. Then hold [H/L] (COMP) while turning the power on. (Radio will show CAL.) Next, rotate the radio around twice, taking about ten seconds per rotation. Finally, push [Enter].

This radio has an extended transmit frequency range (142.000–149.880 MHz and 420.000–470.000 MHz). To enable this, hold $\boxed{\text{SC}}$ and $\boxed{9}$ while turning on the power. This will reset the radio (losing all memories/settings) as well as changing the frequency limits. Repeat the procedure to select normal frequency range.

Adjusting the volume on this radio can affect the knob and vice versa. Use care when moving one that you don't move the other.

Factory reset

To reset the radio hold $\boxed{\text{F}}$ and $\boxed{6}$ (CLR) while turning the radio on. This clears all memories and settings.

Radio Layout

Specs

Receivers Single receiver with priority channel option
Receives 136–174 MHz FM
Transmits 136–174 MHz @ 5 W FM
Antenna connector SMA **M** on radio; needs SMA **F** antenna
Modes FM
Memory Channels 99
Power No DC input on radio; charger has "M"-style barrel (5.5 mm OD, 2.1 mm ID)
    with 12 V DC center positive
Model year 2002

Standard Tasks

Program frequency in the field

This radio was not designed to be field-programmable, and the specifications for the United States version list 148–174 MHz only. However, buried in the firmware is a "dealer mode" which allows programming in the ham bands from the keypad. "Dealer mode" can be disabled via front-panel programming, at which point you will no longer be able to program the radio without software.

1. Turn the radio off.

2. Hold [O] and [**LAMP**] while turning the radio on. Display will show SEL.

3. Press [■].

4. Use the knob to select the memory you want to program (1–99). Press [**PTT**] (does not transmit) to select.

5. You will see the receive frequency if it has already been programmed into the channel. If you see ‑‑‑‑‑‑‑‑ then press [☐] to see the receive frequency.

6. Set receive frequency: use the knob to select. You can hold [**LAMP**] to adjust MHz. Press [●] to toggle between 6.25 kHz step and 2.5 kHz step. Press [**PTT**] (does not transmit) to set.

7. Set receive tone type and value: press [☐] to cycle through OFF (no tone required for receive), QT (CTCSS tone) and DQT (DCS tone). Use the knob to select the desired tone. Press [**PTT**] (does not transmit) to set.

8. You will see the transmit frequency if it has already been programmed into the channel. If you see ‑‑‑‑‑‑‑‑ then press [☐] to see the transmit frequency.

9. Set transmit frequency: use the knob to select. You can hold [**LAMP**] to adjust MHz. Press [●] to toggle between 6.25 kHz step and 2.5 kHz step. Press [**PTT**] (does not transmit) to set.

10. Set transmit tone type and value: press [☐] to cycle through OFF (no tone sent on transmit), QT (CTCSS tone) and DQT (DCS tone). Use the knob to select the desired tone. Press [**PTT**] (does not transmit) to set.

11. You will see `SIG OFF` (no DTMF), `SIG DTMF` (DTMF enabled) or `SIG TTS` (two tone signaling instead of DTMF). Use the knob to adjust, then press [PTT] (does not transmit) to set.

12. You will see `ANI OFF` (no automatic ID) or `ANI ON` (send DTMF ID when PTT is pressed). Use the knob to adjust, then press [PTT] (does not transmit) to set. You probably want this to be off.

13. You will see `SCAN DEL` (don't include this channel on the scan list) or `SCAN ADD` (include this channel on the scan list). Use the knob to adjust, then press [PTT] (does not transmit) to set.

14. You will see `B.C.L.O OFF` (disable busy channel lock-out) or `B.C.L.O ON` (enable busy channel lock-out). Use the knob to adjust, then press [PTT] (does not transmit) to set.

15. You will see `SHFT OFF` (disable clock shift) or `SHFT ON` (enable clock shift). Use the knob to adjust, then press [PTT] (does not transmit) to set. Note that this does not refer to repeater shift.

16. You will see `TXPWR H` (enable high-power transmit) or `TXPWR L` (disable high-power transmit). Use the knob to adjust, then press [PTT] (does not transmit) to set. If you set this to high, you can still reduce power later. If you set this to low, you cannot increase power.

17. You will see `WIDE` (25 kHz bandwidth) or `NARROW` (12.5 kHz bandwidth). Use the knob to adjust, then press [PTT] (does not transmit) to set.

18. You will see `ID` flash and then disappear. Press [PTT] (does not transmit) to move on to the next step.

19. You will see `CH LABEL` flash and then disappear, replaced with the channel name. Press [PTT] (does not transmit) to move on to the next step.

20. Turn the radio off and then back on to exit dealer mode.

Lock/unlock radio

Press the key assigned to keyboard lock to lock/unlock the radio. In the default configuration it is not set, but can be assigned (see "Buttons don't do what I expect" below).

Check repeater input frequency

Depending on how it has been programmed, this radio may have the ability to check the repeater input frequency. The global setting `TARE RE` must be enabled. When enabled, the reverse button is by default programmed as [■]. Reverse mode is indicated with the channel; it is displayed as `rE 5` rather than `CH 5` for channel 5. If you see `tA 5`, then you are in talkaround mode instead, and will transmit and receive on the output frequency.

Change power in the field

Depending on how it has been programmed, this radio may have the ability to adjust output power. The global setting `LO ON` must be enabled. The channel setting `TXPWR H` must be enabled for the channel. By default, the ☐ button will then reduce power from full power (5 W) to low power (1 W). The display will show `Lo` when power has been reduced.

Adjust volume

Turn the power/volume knob to adjust volume.

Adjust squelch

Squelch is set with the global setting `SQL 0` through `SQL 9`. See "Global settings" below for more information. If a key has been programmed for squelch, that can be used instead.

Weird Modes

Buttons don't do what I expect

The buttons are programmable. You can review by going into extended setup. To do this, hold ⟨ ○ ⟩ and [LAMP] while turning the radio on. Display will show `SEL`. Next, press ☐.

Hold [MONITOR] while rotating the knob until you see an entry starting with `K1` (⟨ ● ⟩), `K2` (⟨ ○ ⟩), `K3` (⟨ ■ ⟩) or `K4` (⟨ ☐ ⟩). The value after will indicate which function is currently assigned.

Functions are `OFF` (nothing assigned), `SCAN` (scan), `DIAL` (allow DTMF sending), `TARE` (reverse or talk-around depending on configuration), `LO` (low power), `DCHAR` (display character label for channel), `DFREQ` (display frequency for channel), `DMODE` (switch between display modes), `SADD` (add or delete from scan list), `KLOCK` (enable or disable keyboard lock), `VQT` (enable or disable CTCSS tone adjustment when on a channel), and `SQL` (adjust squelch).

Release [MONI] and use the knob to scroll through the values. Press [PTT] (does not transmit) to set. Turn the radio off then on again to exit extended select mode.

Useful Information

⚠ **WARNING:** It is possible to disable dealer mode from the menus. If you go into dealer mode and choose the setting `MODE OFF`, you will be unable to make changes. Ensure that you always have `MODE ON`.

The radio can enter a number of settings modes. To enter any of them, first turn the radio off, then hold ⟨ ○ ⟩ and [LAMP] while turning the radio on. Next press ⟨ ● ⟩ to enter global settings, ⟨ ○ ⟩ to enter DTMF settings, ⟨ ■ ⟩ to enter channel programming, or ⟨ ☐ ⟩ to enter extended settings.

While in the three settings menus, you can hold **MONITOR** and use the knob to scroll through settings, and press **PTT** (does not transmit) to set. Turn power off and back on to leave the settings modes.

Factory reset

To fully reset the radio, hold ⓞ and **LAMP** while turning the radio on. Simultaneously press ☐ and **PTT** (does not transmit). The red LED will light up as the device is resetting, and will go dark when the device has been reset.

⚠ **WARNING:** Do not turn off the device while it is resetting. This may corrupt memories.

⚠ **WARNING:** Resetting the device may disable dealer mode, preventing you from making further changes without a computer and programming software.

⚠ **WARNING:** In at least some configurations, the Relm RPV599A Plus may permit you to transmit on business or public safety frequencies. Make sure you are in-band when transmitting.

Radio Layout

Radio with Racal branding

Specs

Receivers Single receiver
Receives 136–174 MHz FM
Transmits 136–174 MHz @ 5 W FM, 2 W FM using AA batteries
Antenna connector SMA **M** on radio; needs SMA **F** antenna. Note that this radio
    requires an SF (Motorola-style) SMA connector
Modes FM, P25
Memory Channels 256, 304 on radios with Fire Feature
Power No DC input on radio; 8–14 V DC per service manual
Model year 1998

Standard Tasks

Program frequency in the field

1. Select the memory you want to program using the selector knob. (Depending on how the radio is programmed, you may be able to use the switch to select zones. You may also be able to go to different banks. See "Useful Info" below for details.)

2. Press `Enter`.

3. Press `◇` to scroll up to `PROGRAM`. Press `Enter`.

4. Press `○` to scroll down to `CHANEL`. Press `Enter`.

5. Set mode: press `○` to scroll down to `MODE=`. If that is not already `ANALOG`, press `Enter` to select, press `○` to scroll to `ANALOG`, then press `Enter` to set. Your other choice is `DIGITAL`, which is P25.

6. Set bandwidth: press `○` to scroll down to `B/W`. If that is not already `25 kHz`, press `○` to scroll to `25 kHz`, then press `Enter` to set. Your other choice is 12.5 kHz.

7. Set receive frequency: press `○` to scroll down to `RX=`. Enter the receive frequency using the number pad. (144390 for 144.390 MHz). Press `Enter` to set.

8. Set receive tone type: press `○` to scroll down to `RXSQMD`. Press `Enter` to change. Press `○` to scroll through squelch types. The options are `NONE` (no squelch), `NOISE` (carrier squelch), `TONE` (CTCSS squelch) and `DCS` (DCS squelch). Press `Enter` to set.

9. Set the receive tone or squelch: press `○`. Available types are `SQ=`, `TONE=` or `CODE=`. Select the value using `○` and `◇`. Press `Enter` to set. There are seventeen levels of squelch; the default is 3.

10. Set transmit frequency: press `○` to scroll down to `TX=`. Enter the receive frequency using the number pad. (144390 for 144.390 MHz). Press `Enter` to set.

11. Set transmit tone type: press ⬭ to select `TXSQMD=`. Press **Enter** to change; your options are `NONE`, `CTCSS` and `DCS`. Press **Enter** again to set.

12. Set transmit tone: press ⬭ to select `TON=` (or `CODE=` for DCS). Press **Enter** to change, then adjust with ⬭ and ◇. Press **Enter** to set.

13. Set low power value: press ⬭ to select `LO PWR=` and press **Enter** to set. Use `Circle` and `Diamond` to adjust. You can choose from `0.1 W`, `0.5 W`, `1.0 W`, `2.0 W` and `5.0 W`. Press **Enter** to set.

14. Set high power value: press ⬭ to select `HI PWR=` and press **Enter**. Use ⬭ and ◇ to adjust. You can choose from `0.1 W`, `0.5 W`, `1.0 W`, `2.0 W` and `5.0 W`. Press **Enter** to set.

15. Press ▢ repeatedly to escape out of the menus (or just wait).

Lock/unlock radio

While holding ▢, hold **Enter** for one half second, then release both buttons. Repeat this to cycle through:

- `KEYS DISABLED/SIDE ENABLED`

- `KEYS DISABLED/SIDE DISABLED`

- `KEYS ENABLED/SIDE ENABLED` (unlocked).

Release all buttons for one second to set.

Check repeater input frequency

Although talkaround can be programmed using a computer programmer, this is not possible in the field. You will have to program a separate frequency for this. If a frequency has been programmed with talkaround, press **Enter** and then ⬭ to scroll down to `SELECT`. Press **Enter**. Press ◇ to scroll up to `TKRD=`. Press **Enter**, then press ⬭ to scroll to `ON` or `OFF`. Press **Enter** to set. Press ▢ repeatedly to escape out of the menus. If you are in talkaround mode, the main screen will show `TA`. Otherwise it will show ⋀⋀ to indicate repeater mode.

Change power in the field

Using a PC programmer, one of the side buttons can be programmed to switch between low and high power. If one is not programmed, you can change the level for an individual channel. See "Program freq in the field" for instructions. The display will show `LO` or `HI` to indicate current power level.

Adjust volume

Turn the power/volume knob to adjust volume. The knob has sixteen steps: off, `MUTE` and then fourteen levels of volume.

Adjust squelch

This is set on a per-channel basis. See "Program freq in the field" for instructions. If a side button has been programmed, you can press it to monitor temporarily, hold it for two seconds to monitor continuously, or hold it for four seconds to adjust squelch. (Use ⟨ o ⟩ and ⟨ ◇ ⟩ to adjust.)

Weird Modes

Buttons don't work

First, check to make sure the side buttons are enabled (see the "Lock/unlock radio" section). If that doesn't work, be aware that many of this radio's buttons and switches are programmable using a PC programmer. Depending on how the radio was configured, the side buttons may or may not work, and may do different things.

Radio has open mic, flashing lights and/or sounds

The radio has an emergency mode. This transmits a signal on P25, but on analog channels it just opens the mic. The radio will also display `EMERGENCY`. Depending on the emergency mode programmed, there may also be exciting lights and sounds. Turn the radio off and on again to exit.

Useful Information

⚠ **WARNING:** This radio uses an SF (Motorola-style) SMA connector. This requires that the female side of the connector have a center conductor that extends to the end of the connector. Standard SMA connectors do not do this. Be sure your antenna is making contact with the radio's connector.

Standard connector won't work   SF (Motorola-style) SMA will work

In the United States, programming a frequency with NOAA weather radio is a quick way to test if the antenna is making connection.

It is possible to set the radio to display frequency rather than channel name. To do this, press ⟨**Enter**⟩, then use ⟨ ◇ ⟩ to scroll to `PROGRAM`. Press ⟨**Enter**⟩. Ensure that `GLOBAL` is selected. (If not, use ⟨ o ⟩ to scroll to it.) Press ⟨**Enter**⟩. Press ⟨ o ⟩ to scroll to `DISPLY`. Press ⟨**Enter**⟩. Press ⟨ o ⟩ to scroll between `ALPHA` (channel name) and `NUMBER` (frequency). Press ⟨**Enter**⟩ to set. Press ⟨ □ ⟩ repeatedly to escape out of the menu system. The default password is 000000.

When a channel has a square box around it, that means it's in the scan list. Press ⟨ o ⟩ to remove or ⟨ ◇ ⟩ to add it.

If your radio has the Fire Feature, you can switch to bank 5. Press **Enter**, then scroll to `SELECT` using ⟨ O ⟩. Press **Enter**. Scroll to `ZONE` using ⟨ O ⟩. Press **Enter**. Choose either `1 BANK 01` or `5 BANK 05`. Press **Enter**. Press ⟨ □ ⟩ to escape out of the menu system.

Some radios have a `OTAR` (over-the-air re-keying) mode that shows up when you first press **Enter**. Use ⟨ O ⟩ to get to the regular menu system.

No reset procedure

This radio does not have a way to reset from the keypad.

⚠ **WARNING:** In at least some configurations, the Thales 25 may permit you to transmit on business or public safety frequencies. Make sure you are in-band when transmitting.

Radio Layout

Specs

Receivers Two independent receivers, simultaneous receive
Receives 136–174 MHz, 400–470 MHz
Transmits 136–174 MHz FM @ 5 W, 400–470 MHz FM @ 5 W
Antenna connector SMA **M** on radio; needs SMA **F** antenna
Modes FM
Memory Channels 199
Power 6.6–8.4 V DC per specification, No DC input on radio
Model year 2014

Standard Tasks

Program frequency in the field

1. If you aren't already in VFO mode, press [*]. The screen will show `Ch nnn` if you are not in VFO mode, where nnn is 001–199.

2. This radio doesn't allow you to overwrite memories. You will need to delete first if you want to write to a memory that has data in it. To delete: press [MENU] [3] [3] [MENU] (`Del Ch`). Use the knob or keypad to select the channel to delete (001–199). Then press [MENU] [A/B].

3. Set frequency: use the keypad (144390 for 144.390 MHz).

4. Set transmit tone type and value: press [MENU] [8] [MENU] (`Tx Tone`) then scroll to the correct CTCSS tone frequency using the center selection knob, press [MENU] [A/B]. If you see DCS tones or `OFF`, press [#] until you see CTCSS tones.

5. Set receive tone type and value: press [MENU] [7] [MENU] (`Rx Tone`) then scroll to the correct CTCSS tone frequency using the center selection knob, press [MENU] [A/B]. If you see DCS tones or `OFF`, press [#] until you see CTCSS tones.

6. Set repeater shift: press [MENU] [6] [MENU] (`Rpt Dir`) and scroll to the correct repeater shift (`+`, `–` or `OFF`). Press [MENU] [A/B].

7. If you need to change the repeater offset frequency: press [MENU] [5] [MENU] (`Offset`) and scroll to the correct repeater offset, then press [MENU] [A/B].

8. Set transmit power: press [MENU] [2] [MENU] (`TXPower`) and scroll to correct power level (`LOW` (0.5 W), `MID` (3 W VHF, 2 W UHF) or `HIGH` (5 W), then press [MENU] [A/B].

9. Write to a memory: press [MENU] [3] [2] [MENU] (`Save Ch`). Use the knob or keypad to select the channel to write (001–199) then press [MENU] [A/B].

10. Go to memory mode: press [*]

11. Select the memory you just wrote: use the knob.

Lock/unlock radio

Hold [#] for three seconds to lock/unlock.

Check repeater input frequency

Hold [*] for two seconds. The R indicator will show you're in reverse mode. Hold [*] again to return to normal mode.

Change power in the field

To set transmit power, press [MENU] [2] [MENU] (TXPower) and scroll to correct power level (LOW (0.5 W), MID (3 W VHF, 2 W UHF) or HIGH (5 W), then press [MENU] [A/B].

Adjust volume

Rotate the right power/volume knob to adjust volume.

Adjust squelch

Press [MENU] [1] [MENU] (SqlLevel) then scroll to the correct squelch level (0–9). Then press [MENU] [A/B].

Weird Modes

Keys don't work and display shows FM

You are in FM radio mode. This is typically entered via a side button; pressing that button again will leave the mode. If that doesn't work, try [MENU] [1] [9] (FmRadio) or [MENU] [2] [0] (Fm Mon). Press [MENU], then use the knob to change to OFF. Then press [MENU] [A/B].

Side buttons don't behave as expected

This radio has two buttons on the left hand side below [PTT]. They can be programmed for a number of different functions.

Can't leave channel (memory) mode

This radio is often shipped with VFO mode turned off. If that's the case, you will need to modify the programming with a computer/radio programming cable to turn VFO mode on. This cannot be changed from the front panel.

Useful Information

To switch between monitoring one frequency and monitoring two frequencies, use [MENU] [2] [2] [MENU] (SubDisp) then scroll to either Dual or Single, then press [MENU] [A/B].

Factory reset

To reset the radio, hold [A/B] while turning the radio on. Use the knob to choose ALL to reset everything. Press [MENU].

VFO reset

To reset the radio, hold [A/B] while turning the radio on. Use the knob to choose VFO to reset VFO-related settings only). Press [MENU].

⚠ **WARNING:** In at least some configurations, the Tera TR-590 may permit you to transmit on business or public safety frequencies. Make sure you are in-band when transmitting.

Tytera TYT MD-2017

Radio Layout

Power/
Volume

PTT

▲

▼

Trackball

MENU

*

Specs

Receivers Single receiver
Receives 136–174 MHz FM/DMR, 400–480 MHz FM/DMR
Transmits 136–174 MHz FM/DMR @ 5 W , 400–480 MHz FM/DMR @ 5 W
Antenna connector SMA F on radio; needs SMA M antenna
Modes FM, DMR
Memory Channels 3000
Power 7.4 V DC, No DC input on radio
Model year 2017

Standard Tasks

Program frequency in the field

⚠ **WARNING:** This radio can be configured to prevent front panel programming, or to allow it only with a password. If it's configured this way, you will not be able to program the radio.

1. Enter programming mode: press `MENU` `▲` `MENU` to select the `Utilities` menu. Then press `▲` `MENU` to select the menu item `Program Radio`. Enter the password if required (many code plugs use 99999999 or 00000000 as the eight digit password), then press `MENU`.

2. Press `▼` or use the trackball to scroll to menu 6, `Add CH`. Press `MENU`.

3. Make sure item 1, `Analog CH` is selected. Push `MENU`.

4. Enter the channel name: (display shows `Enter CH Name`). Use the trackball to adjust the cursor position, and the keypad to enter a name. Press `MENU`.

5. Set receive frequency: (display shows `Rx Frequency`). Use the trackball to back up and then the keypad to enter (144390 for 144.390 MHz). Press `MENU`.

6. Set transmit frequency: (display shows `Tx Frequency`). Use the trackball to back up and then the keypad to enter (144390 for 144.390 MHz). Note that you set receive and transmit frequency rather than an offset. Press `MENU`. The channel is added.

7. Add the channel to a new zone: press `MENU` and select `Zone` with the trackball. Press `MENU`.

8. Select `New Zone` with the trackball. You will be prompted for a zone name. Use the trackball to adjust cursor position and the keypad to enter a zone name. Press `MENU`.

9. Select the new zone: press `MENU` and use the trackball to select `ZONE`. Press `MENU`.

10. Use the trackball to choose `ZoneList`. Press `MENU`.

11. Use the trackball to choose `Add CH`. Press `MENU`.

12. Select `CH A` and press `MENU`. You will see `Add Channel Successful`. Your channel is now added to the zone.

13. Turn the zone on: press `MENU` then select `Zone`, press `MENU` and select `ZoneList`. Press `MENU`. Scroll to your zone and press `MENU`. Select menu item 1, `On`. Press `MENU`.

14. Your channel is now set in your zone. Make sure your channel is selected, then press `MENU`. Use the trackball to scroll to `Utilities` and press `MENU`.

15. Use the trackball to scroll to `Program Radio` and press `MENU`.

16. Scroll to `Edit Channel` and press `MENU`.

17. You can now scroll to `CTC/DCS` and press `MENU`.

18. Scroll to `T CTC` and press `MENU`.

19. Use the trackball up/down to select the tone you want. Press `MENU` when done.

20. Press `MENU` then `▲` `MENU` to select `Utilities`. Press `MENU` to choose `Radio Settings`. Press `▼` twice to select `Power`. Press `MENU` to confirm. Use `▲` `▼` to select the power level you want (`High` is 5 W, `Low` is 1 W), then press `MENU` to set.

Lock/unlock radio

To lock the radio, press `MENU` and go to the `Utilities` menu. Press `MENU`. Scroll to menu 7, `Keypad Lock`. Press `MENU`. Use the trackball to choose from `Manual` (no lock), or a time to lock (`5`, `10` or `15` seconds). Press `MENU`. Wait for the radio to lock.

To unlock the radio, press `MENU` `*`.

Check repeater input frequency

This radio has no option to listen to the repeater input frequency. You will have to program a separate memory with the repeater input frequency to do this.

Change power in the field

To set transmit power, Press `MENU` then `▲` `MENU` to select `Utilities`. Press `MENU` to choose `Radio Settings`. Press `▼` twice to select `Power`. Press `MENU` to confirm. Use `▲` `▼` to select the power level you want (`High` is 5 W, `Low` is 1 W), then press `MENU` to set.

Adjust volume

Rotate the power/volume knob to adjust the volume.

Adjust squelch

Press **MENU** then **▲** then **MENU** to select `Utilities`. Then press **MENU** to choose `Radio Settings`. Press **▼** four times to select `Squelch`. Press **MENU** to confirm. Use **▲** or **▼** to select the squelch you want (`Tight` or `Normal`). Press **MENU** to set.

Weird Modes

Radio doesn't transmit

This radio has a transmit inhibit feature. To disable it, press **LEFT MENU** then **▲** then **MENU** to select `Utilities`. Then press **MENU** to choose `Radio Settings`. Press **▼** to select `Tones/Alerts`, then press **MENU** to confirm. Press **▼** to select `Talk Permit`, and press **MENU**. Use **▲** or **▼** to choose `Turn On` (user can transmit) or `Turn Off` (user can't transmit). Press **MENU** to set.

Radio stops receiving and/or transmitting after receiving a signal

This radio has a remote kill feature that can prevent the radio from transmitting or receiving when it receives a properly-formatted transmission. Most of the time, this requires programming software to revive the radio. A remote activation feature exists, but that requires sending a code from another TYT MD-2017.

Useful Information

The radio's memories are divided into zones (similar to banks), with a maximum of sixteen memories per bank. To change zones, from the root menu press **MENU** then **▼** to select `Zone`. Press **MENU** to confirm. Use the trackball to select `ZoneList` and press **MENU**. Select your zone and press **MENU**. Select `On` and press **MENU** to confirm. To change memories within a zone, use the trackball.

The **MENU** button on the radio has a top function and a bottom function. You need to press the bottom of the button to get the **MENU** function.

This radio has an unusual antenna connector, which is wider than normal (although still SMA).

MD-2017 antenna

On at least some versions of the radio, pressure on the antenna could snap the SMA connector, requiring replacement. Aftermarket antennas may put additional stress on the radio's SMA connector.

The radio menu system has a ten-second timeout. You will need to act quickly once you have entered the menu.

This radio has the ability to record received audio.

Although this radio is dual-band, only one band can be active at a time.

This radio has frequent firmware updates (and sometimes has third-party firmware loaded as well). These instructions are for stock firmware V003.033.

No reset procedure

The radio does not have a way to reset it without programming a new codeplug. (The radio does reset when a new codeplug is uploaded.)

⚠ **WARNING:** In at least some configurations, the Tytera TYT MD-2017 may permit you to transmit on business or public safety frequencies. Make sure you are in-band when transmitting.

Tytera TYT MD-380

Radio Layout

Specs

Receivers Single receiver
Receives One of 136–174 MHz FM/DMR or 400–480 MHz FM/DMR
Transmits One of 136–174 MHz FM/DMR @ 5 W or 400–480 MHz FM/DMR @ 5 W
Antenna connector SMA F on radio; needs SMA M antenna
Modes FM, DMR
Memory Channels 1000
Power No DC input on radio; 12.5 V DC, 5.5 mm OD, 2.1 mm ID, center positive on charger base
Model year 2015

Standard Tasks

Program frequency in the field

This radio does not allow you to add new frequencies in the field. However, you can modify existing frequencies that have been programmed. Many codeplugs have an "FPP" zone intended for this purpose.

⚠ **WARNING:** This radio can be configured to prevent front panel programming, or to allow it only with a password. If it's configured this way, you will not be able to program the radio.

1. On this radio, you select the destination memory first. To select: use the knob. Note that you can't convert an FM memory into a DMR memory or vice versa, and you can't change talkgroup. You may need to change zones; see "Useful Info" below.

2. To enter programming mode, start by pressing `MenuL` `▲` `MenuL` to select the `Utilities` menu. Then press `▲` `MenuL` to select the menu item `Program Radio`. Enter the password (many code plugs use 99999999 or 00000000 as the eight digit password), then press `MenuL`.

3. Set receive frequency: press `MenuL` `MenuL`.

4. Press `▲` repeatedly to delete previous frequency.

5. Enter the desired frequency on the keypad (445925 for 445.925 MHz). Press `MenuL` to set.

6. Set transmit frequency: press `▼` `MenuL` `MenuL`.

7. Press `▲` repeatedly to delete previous frequency.

8. Enter the desired frequency on the keypad (445925 for 445.925 MHz). Press `MenuL` to set.

9. Set receive tone type and value: press `▼` `▼` `▼` `MenuL`. If you can't get into that menu, you're on a DMR memory and not an analog memory. Press `MenuL` to go into the `R CTC` (receive CTCSS tone) menu. Press `MenuL` to change, use `▲` `▼` to select, then press `MenuL` to set. You can change `R DCS` instead if you use DCS tones.

10. Set transmit tone type and value: press [▼] [▼] [MenuL] to set T CTC (transmit CTCSS tone). Press [MenuL] to change, use [▲]/[▼] to select, then press [MenuL] to set. You can change R DCS instead if you use DCS tones.

11. Press [MenuR] repeatedly to exit the menu system.

12. Set transmit power: press [MenuL] then [▲] then [MenuL] to select Utilities. Press [MenuL] to choose Radio Settings. Press [▼] twice to select Power. Press [MenuL] to confirm. Use [▲]/[▼] to select the power level you want (High is 5 W, Low is 1 W), then press [MenuL] to set.

Lock/unlock radio

Hold [*] for two seconds to lock. Press [MenuL] [*] to unlock when locked. (Note that the radio can be configured to auto-lock by going into the Radio Settings menu with Keypad Lock.)

Check repeater input frequency

This radio has no option to listen to the repeater input frequency. You will have to program a separate memory with the repeater input frequency to do this.

Change power in the field

To set transmit power, press [MenuL] then [▲] then [MenuL] to select Utilities. Then press [MenuL] to choose Radio Settings. Press [▼] twice to select Power. Press [MenuL] to confirm. Use [▲]/[▼] to select the power level you want (High is 5 W, Low is 1 W), then press [MenuL] to set.

Adjust volume

Rotate the power/volume knob to adjust the volume.

Adjust squelch

Press [MenuL] then [▲] then [MenuL] to select Utilities. Then press [MenuL] to choose Radio Settings. Press [▼] four times to select Squelch. Press [MenuL] to confirm. Use [▲]/[▼] to select the squelch you want (Tight or Normal). Press [MenuL] to set.

Weird Modes

Radio doesn't transmit

This radio has a transmit inhibit feature. To disable it, press [MenuL] then [▲] then [MenuL] to select Utilities. Then press [MenuL] to choose Radio Settings. Press [▼] to select Tones/Alerts, then press [MenuL] to confirm. Press [▼] to select `Talk Permit`, and press `MenuL`. Use `▲` `▼` to choose `Turn On` (user can transmit) or `Turn Off` (user can't transmit). Press `MenuL` to set.

Radio stops receiving and/or transmitting after receiving a signal

This radio has a remote kill feature that can prevent the radio from transmitting or receiving when it receives a properly-formatted transmission. Most of the time, this requires programming software to revive the radio. A remote activation feature exists, but that requires sending a code from another TYT MD-380.

Radio menus are in Chinese

To reset the radio to English, first press `MenuL`, `▲`, `MenuL`, then `MenuL`. Press `▼` seven times (until you are on `8`) then press `MenuL`. Press `▲` to select `English` and press `MenuL` to set.

Useful Information

The radio's memories are divided into zones (similar to banks), with a maximum of sixteen memories per bank. To change zones, from the root menu press `MenuL` then `▼` to select `Zone`. Press `MenuL` to confirm. Press `▲` `▼` to select a zone, then press `MenuL` to confirm. To change memories within a zone, use the knob.

The radio menu system has a ten-second timeout. You will need to act quickly once you have entered the menu. In particular, you will need to re-enter the programming password if you time out.

No reset procedure

The radio does not have a way to reset it without programming a new codeplug. (The radio does reset when a new codeplug is uploaded.)

⚠ **WARNING:** In at least some configurations, the Tytera TYT MD-380 may permit you to transmit on business or public safety frequencies. Make sure you are in-band when transmitting.

Radio Layout

Power/
Volume

PTT

F

▲

▼

Exit

*   #

Specs

Receivers Single receiver, dual watch (first to break squelch wins)
Receives Single band: one of 136–174 MHz FM or 245–246 MHz FM or 400–470 MHz FM or 350–390 MHz FM or 465–520 MHz FM
Transmits Single band: one of 136–174 MHz FM @ 5 W or 245–246 MHz FM @ 5 W or 400–470 MHz FM @ 4 W or 350–390 MHz FM @ 4 W or 465–520 MHz FM @ 4 W
Antenna connector SMA F on radio; needs SMA M antenna
Modes FM
Memory Channels 128
Power No DC input on radio; 12 V DC, 5.5 mm OD, 2.1 mm ID, **center negative** on charger base
Model year 2010

Standard Tasks

Program frequency in the field

1. If you aren't already in VFO mode, press **EXIT**. The channel number is not shown in VFO mode.

2. If needed, set step size: press **F** **2** **9** **F**. The display will show STEP - 29 and the current value. Use **▲** **▼** to change, then press **F** **EXIT** **EXIT** to save.

3. Set frequency: use the keypad (144390 for 144.390 MHz).

4. Set transmit tone type: press **F** **2** **7** **F**. The display will show T-CDC -- 27. Then press ***** repeatedly until you see 67.0 (for CTCSS) or DO23N (for DCS). Use **▲** **▼** to set the value to your choice. Note that menu 27 sets the transmit tone; you can use menu 26 (R-CDC - 26) for receive tone and menu 25 (C-CDC - 25) for both transmit and receive tone. Once you have set the tone, press **F** **EXIT** **EXIT** to save.

5. If you need to change the repeater offset frequency: press **F** **2** **3** **F**. The display will show OFFSET- 23 and the current value. Press **▲** **▼** to adjust, or enter the offset directly on the number pad. (00600 is the standard 600 kHz offset, which is displayed as 0.600.) Press **F** **EXIT** **EXIT** to save.

6. Set repeater shift: press **F** **2** **8** **F**. The display will show S-D - 28 and the current value. Press **▲** **▼** to adjust among OFF, - and +. Press **F** **EXIT** **EXIT** to save.

7. Set transmit power: press **F** **4** **F**. The display will show POW - 04 and the current value. Press **▲** **▼** to adjust between HIGH (5 W VHF or 4 W UHF) and LOW (1 W VHF or 0.5 W UHF). Press **F** **EXIT** **EXIT** to save.

8. Write to a memory: press **F** **EXIT**. Use **▲** **▼** to select the memory to write (memory will flash if it already has data in it). Press **EXIT** to store. You will automatically enter memory mode after writing the memory.

Lock/unlock radio

Hold [*] for two seconds to lock/unlock.

Check repeater input frequency

Hold [#] for two seconds to switch to reverse frequency. The display will show R when in reverse mode. Hold [#] for two seconds to leave reverse frequency mode.

Change power in the field

To set transmit power, press [F][4][F]. The display will show POW - 04 and the current value. Press [▲][▼] to adjust between HIGH (5 W VHF or 4 W UHF) and LOW (1 W VHF or 0.5 W UHF). Press [F][EXIT][EXIT] to save.

Adjust volume

Rotate the power/volume knob to adjust the volume.

Adjust squelch

Press [F][5][F] to set power level. The display will show SQL - 05 and the current value. Press [▲][▼] to adjust from 0–9 (default is 5). Press [F][EXIT][EXIT] to save.

Weird Modes

Transmit audio is distorted

This radio has a built-in scrambler which may not be legal for amateur radio use. To check, press Press [F][3][4][F]. If the radio displays APRO - 34 then it has the scrambler. Use [▲][▼] to set it to OFF, then press [F][EXIT][EXIT] to save.

Radio stops receiving and/or transmitting after receiving a signal

This radio has a remote kill feature that can prevent the radio from transmitting or receiving when it receives a properly-formatted transmission. Most of the time, this requires programming software to revive the radio.

Radio doesn't enter VFO mode

The radio has a "Channel Mode" that prevents direct frequency entry. To enable/disable this mode, hold [▲] while turning the radio on.

Useful Information

The password for the AGING ? reset is often 5858.

Some users have reported problems after using the programming software. The radio goes into a mode where its output power is limited to 1 W. The dealer programming software may be able to reset this.

Factory reset

Hold [F] while turning the radio on, then press [F]. The display will show `VFO ?`. Press [▲][▼] to select `FULL ?` (resets everything). If you decide you don't want to reset, turn the radio off and back on. Press [F] to perform the reset.

VFO reset

Hold [F] while turning the radio on, then press [F]. The display will show `VFO ?`. Press [▲][▼] to select `FULL ?` (resets everything). If you decide you don't want to reset, turn the radio off and back on. Press [F] to perform the reset.

Aging Reset

This radio has a third kind of reset. The meaning of this is not known, but it may have something to do with allowing/preventing out-of-band transmissions. Hold [F] while turning the radio on, then press [F]. The display will show `VFO ?`. Press [▲][▼] to select `AGING ?`. If you decide you don't want to reset, turn the radio off and back on. Press [F] to perform the reset.

⚠ **WARNING:** In at least some configurations, the Tytera TYT TH-F8 may permit you to transmit on business or public safety frequencies. Make sure you are in-band when transmitting.

Tytera TYT TH-UV88

Radio Layout

Specs

Receivers Single receiver, dual watch (first to break squelch wins)
Receives 136–174 MHz FM and 400–480 MHz FM
Transmits 136–174 MHz @ 5 W FM and 400–480 MHz @ 5 W FM
Antenna connector SMA F on radio; needs SMA M antenna
Modes FM
Memory Channels 200
Power No DC input on radio; operating voltage is 7.4 V DC
Model year 2020

Standard Tasks

Program frequency in the field

1. If you aren't already in VFO mode, press `#`. The screen will show `VFO`.

2. Set frequency: use the keypad (144390 for 144.390 MHz).

3. Set transmit tone type and value: press `F` `2` `5` `F` (`T-CTC`) then use `▲`/`▼` to select CTCSS tone frequency (or `OFF`). Press `F` `A/B` to set.

4. Set receive tone type and value: press `F` `2` `4` `F` (`R-CTC`) then use `▲`/`▼` to select CTCSS tone frequency (or `OFF`). Press `F` `A/B` to set.

5. Set repeater shift: press `F` `2` `6` `F` (`S-D`) and use `▲`/`▼` to select repeater shift from `OFF` (simplex), `+` (positive shift), or `-` (negative shift). Press `F` `A/B` to set.

6. If you need to change the repeater offset frequency: press `F` `2` `2` `F` (`OFFSET`) and use `▲`/`▼` to choose the repeater offset (00600 for 600 kHz, 05000 for 5 MHz), then press `F` `A/B` to set.

7. Set transmit power: press `F` `4` `F` (`TXP`) then use `▲`/`▼` to select power level: `LOW` (1 W), `MID` (2.5 W) or `HIGH` (5 W). Press `F` `A/B` to set.

8. Write to a memory: hold `F` then use `▲`/`▼` to select the memory to write. Press `#` to write.

9. Go to memory mode: press `#`.

10. Select the memory you just wrote: use `▲`/`▼`.

Lock/unlock radio

Hold `*` for two seconds to lock/unlock.

Check repeater input frequency

This radio has no option to listen to the repeater input frequency. You will have to program a separate memory with the repeater input frequency to do this.

Change power in the field

To set transmit power, press F 4 F (TXP) then use ▲ / ▼ to select power level: LOW (1 W), MID (2.5 W) or HIGH (5 W). Press F A/B to set.

Adjust volume

Rotate the power/volume knob to adjust volume.

Adjust squelch

Press F 5 F (SQL) then use ▲ / ▼ to select squelch level (0–9). Press F A/B to set.

Weird Modes

Can't enter VFO mode

This radio has two types of channel-only mode that can prevent users from entering frequencies. The first is controlled by menu 33, DIS MD. If it is set to CH, the VFO is not available. Make sure it is set to MR (memory) or FRE (frequency).

If that doesn't work, the radio can be configured to prevent users from entering frequencies entirely regardless of display mode. To disable (or enable) this setting, turn the radio off. Hold LIGHT and MONI and turn the radio on. Release all buttons. Frustratingly, after you do this the radio may always start up with the keylock on. If that's the case, you will need to unlock it before you use the radio. Performing a factory reset may clear this.

Radio makes alarm sound

Pressing F # will send an alarm signal over audio and on the currently selected frequency. This **does transmit.**

Useful Information

If you wait too long after pressing F , the menu mode on the radio will time out. You need to act quickly.

This radio does not choose offset based on frequency. You will have to push ▲ / ▼ a lot to select the offset if you want to switch between 600 kHz and 5 MHz. You can, however, store different offsets on the A and B sides of the radio. This lets you set an offset of 600 kHz on A and 5 MHz on B—you can then program all the 2 m frequencies on the A side and all the 70 cm frequencies on the B side.

Factory reset

To reset the radio, press F 3 5 F . Press ▼ to choose ALL if it isn't already selected. Press F . The radio will reset with no prompt.

Memory reset

To reset all the memories as well as the settings, press [F][3][5][F] Press [▼] to choose FULL if it isn't already selected. Press [F]. The radio will reset with no prompt.

Settings reset

To reset the settings of the VFO only, press [F][3][5][F] Press [▼] to choose VFO if it isn't already selected. Press [F]. The radio will reset with no prompt.

⚠ **WARNING:** In at least some configurations, the TYT TH–UV88 may permit you to transmit on business or public safety frequencies. Make sure you are in-band when transmitting.

Wouxun KG-UVD1P, -UV2D, -UV3D

Radio Layout

KG–UV2D model

Specs

Receivers Single receiver, dual watch (first to break squelch wins)

Receives 76–108 MHz WFM, 136–174 MHz FM and 420–520 MHz FM, 136–174 MHz and 350–470 MHz FM, or 136–174 MHz and 216–280 MHz FM (depending on model)

Transmits 136–174 MHz @ 5 W FM and 420–520 MHz @ 4 W FM, 136–174 MHz @ 5 W FM and 350–470 MHz @ 4 W FM, or 136–174 MHz @ 5 W FM and 216–280 MHz @ 5 W FM (depending on model)

Antenna connector SMA **M** on radio; needs SMA **F** antenna

Modes FM

Memory Channels 128

Power No DC input on radio; charger has "M"-style barrel (5.5 mm OD, 2.1 mm ID) with 12 V DC center positive

Model year 2010

Standard Tasks

Program frequency in the field

1. This radio doesn't allow you to overwrite memories. You will need to delete first if you want to write to a memory that has data in it. To delete: press [MENU] [2] [8] [MENU] (DEL-CH) *XXX* where *XXX* is the channel (001–128). Press [MENU].

2. If you aren't already in VFO mode, press [MENU] then [TDR]. The screen will show small channel numbers on the right hand side if you are not in VFO mode.

3. Set frequency: use the keypad (144390 for 144.390 MHz).

4. Set transmit tone: press [MENU] [1] [6] [MENU] (T-CTC) then use the knob to select CTCSS tone frequency (or off) using the knob. Press [MENU] [EXIT].

5. Set repeater shift: press [MENU] [2] [4] [MENU] (SFT-D) and use the knob to select repeater shift. Press [MENU] [EXIT].

6. If you need to change the repeater offset frequency: press [MENU] [2] [3] [MENU] (OFFSET) and use the keypad to adjust repeater offset (000600 for 600 Hz, 005000 for 5 MHz). Press [MENU] [EXIT].

7. Set transmit power: press [MENU] [4] [MENU] and use the knob to select power level: LOW (1 W) or HIGH (5 W VHF, 4 W UHF)). Press [MENU] [EXIT].

8. Write to a memory: press [MENU] [2] [7] [MENU] (MEM-CH) and enter channel to write *XXX* (001–128). Press [MENU] [EXIT].

9. Go to memory mode: press [MENU] [TDR] to enter channel (memory) mode.

10. Select the memory you just wrote: use the knob.

Lock/unlock radio

Hold [#] for five seconds to lock/unlock.

Check repeater input frequency

Press [*] for less than one second and release. R indicator will show you're in reverse mode. Press and release [*] again to return to normal mode.

Change power in the field

To set transmit power, press [MENU][4][MENU] and use the knob to select power level: LOW (1 W) or HIGH (5 W VHF, 4 W UHF)). Then press [MENU][EXIT].

Adjust volume

Rotate the right power/volume knob to adjust volume.

Adjust squelch

Press [MENU][2][MENU] (SQL-LE) then use the knob to select the squelch level, then press [MENU][EXIT].

Weird Modes

Radio displays countdown when locking

This is normal for some models; ignore it.

Entering frequency "doesn't take"

You are probably in channel (memory) mode with frequency display turned on, instead of VFO mode. Switch to VFO mode.

Useful Information

Menu numbers are for the KG-UV2D. Other models may have them in different positions.

If you wait too long after pressing [MENU], the menu mode will time out. You need to press quickly.

This radio comes in three models: two for 2 m/70 cm and one for 2 m/1.25 m.

There is considerable battery drain when the radio is turned off. Disconnect the battery (or place a piece of paper between the battery terminals and the radio) to preserve the battery when not in use.

This radio is sometimes shipped in a mode that prevents entering new frequencies. If that's the case, you can enable "dealer mode." To enter this, hold [8] while turning the radio on, then enter 268160. This switches channel display (menu 21, CH-MDF) to CH instead of NAME or FREQ, so you will have to set it back using the menu. Dealer mode will be deactivated when you turn off the radio.

Factory reset

To reset the radio, press [MENU] [2] [9] [MENU]. Press [▼] to choose ALL. Press [MENU]. You will see SURE? in the display. (If the device has a password set, you will see - - - - - - and you must enter the correct six-digit password to continue.) Push [MENU] to reset.

VFO reset

To reset the radio, press [MENU] [2] [9] [MENU]. Press [▼] to choose VFO if it isn't already selected. Press [MENU]. You will see SURE? in the display. (If the device has a password set, you will see - - - - - - and you must enter the correct six-digit password to continue.) Push [MENU] to reset.

⚠ WARNING: In at least some configurations, the Wouxun KG-UVD1P, KG-UV2D and KG-UV3D may permit you to transmit on business or public safety frequencies. Make sure you are in-band when transmitting.

Wouxun KG-UV3X Pro

Radio Layout

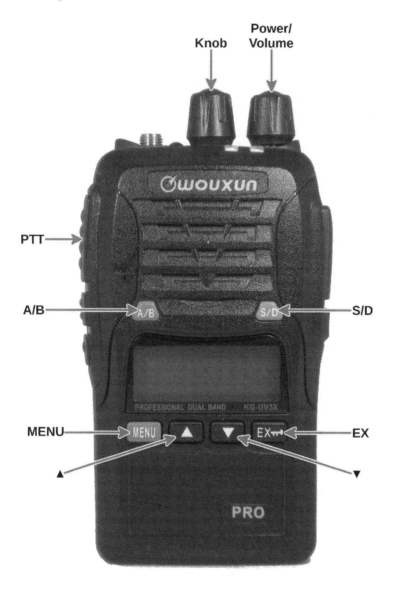

Specs

Receivers Single receiver, dual watch (first to break squelch wins)
Receives 76–108 MHz WFM, 136–174 MHz FM and 375–512 MHz FM
Transmits 136–174 MHz @ 5 W FM and 375–512 MHz @ 4 W FM
Antenna connector SMA F on radio; needs SMA M antenna
Modes FM
Memory Channels 128
Power No DC input on radio; charger has "M"-style barrel (5.5 mm OD, 2.1 mm ID) with 12 V DC center positive
Model year 2012

Standard Tasks

Program frequency in the field

1. This radio doesn't allow you to overwrite memories. You will need to delete first if you want to write to a memory that has data in it. To delete: select the memory to be deleted using the knob. Then press [MENU] and use the knob to select menu 28, DEL-CH. Press [MENU] [MENU] to delete. Press [EX].

2. If you aren't already in VFO mode, press [MENU] then [S/D]. The screen may show small channel numbers on the right hand side if you are not in VFO mode.

3. Set frequency: use the knob.

4. Set transmit tone type and value: press [MENU] then use the knob to select menu 16, T-CTC. Press [MENU] and use the knob to select the frequency. Press [MENU] [EX]. You can also adjust R-CTC (menu 15), R-DCS (menu 17) and T-DCS (menu 18) if desired.

5. Set repeater shift: press [MENU] then use the knob to select menu 24, SFT-D. Press [MENU] to alter, and use the knob to select from OFF, + and -. Press [MENU] [EX].

6. If you need to change the repeater offset frequency: press [MENU] then use the knob to select menu 23, OFFSET. Press [MENU] to alter, and use the knob to adjust. Press [MENU] [EX].

7. Set transmit power: press [MENU] then use the knob to select menu 4, TXP. Press [MENU] to alter, and use the knob to select from HIGH (5 W VHF, 4 W UHF) and LOW (1 W). Press [MENU] [EX].

8. Write to a memory: press [MENU] then use the knob to select menu 27, MEM-CH. Press [MENU] then use the knob to select the desired channel. You will not be able to write into any channel that already has something stored in it. Press [MENU] [EX].

9. Go to memory mode: press [MENU] [S/D].

10. Select the memory you just wrote: use the knob.

Lock/unlock radio

Hold [EX] for two seconds to lock/unlock.

Check repeater input frequency

There is no way to check the input except to program another memory.

Change power in the field

To set transmit power, press [MENU] then use the knob to select menu 4, TXP. Press [MENU] to alter, and use the knob to select from HIGH (5 W VHF, 4 W UHF) and LOW (1 W). Press [MENU] [EX].

Adjust volume

Rotate the right power/volume knob to adjust volume.

Adjust squelch

Press [MENU] then use the knob to select menu 2, SQL-LE. Press [MENU] then use the knob to select the squelch level (0–9). Press [MENU] [EX].

Weird Modes

Can't switch to VFO or memory (channel) mode

This radio is shipped as a Part 90 (commercial) radio. You will first have to enable front panel programming using a computer, programming cable and software before you can do anything from the keypad. In addition, a factory reset of the radio will disable front panel programming.

Entering frequency "doesn't take"

You are probably in channel (memory) mode with frequency display turned on, instead of VFO mode. Switch to VFO mode.

Useful Information

If you wait too long after pressing [MENU], the menu mode will time out. You need to press quickly.

This radio sometimes ships with mode switch password 123456 and reset password 654321. If the password is 000000, you won't be prompted.

You can use [▲] [▼] in place of the knob.

There is considerable battery drain when the radio is turned off. Disconnect the battery (or place a piece of paper between the battery terminals and the radio) to preserve the battery when not in use.

Factory reset

To reset the radio, press [MENU] then use the knob to select menu 29, RESET. Press [MENU]. Use the knob to choose ALL. Press [MENU]. You will see SURE? in the display. (If the device has a password set, you will see − − − − − − and you must enter the correct six-digit password to continue.) Push [MENU] to reset.

⚠ **WARNING:** A factory reset will require you to use a computer, programming cable and software to re-activate front panel programming.

VFO reset

To reset the radio, press [MENU] then use the knob to select menu 29, RESET. Press [MENU]. Use the knob to choose ALL. Press [MENU]. You will see SURE? in the display. (If the device has a password set, you will see − − − − − − and you must enter the correct six-digit password to continue.) Push [MENU] to reset.

⚠ **WARNING:** In at least some configurations, the Wouxun KG-UV3X Pro may permit you to transmit on business or public safety frequencies. Make sure you are in-band when transmitting.

Wouxun KG-UV6D, KG-UV6X

Radio Layout

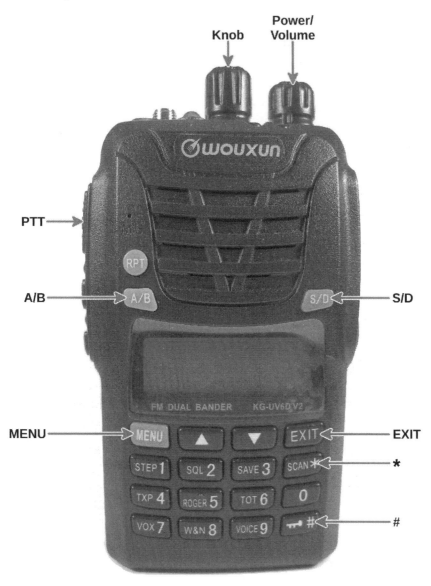

Specs

Receivers Single receiver, dual watch (first to break squelch wins)

Receives 76–108 MHz WFM, 136–174 MHz FM and 420–520 MHz FM, 136–174 MHz and 350–470 MHz FM, or 136–174 MHz and 216–280 MHz FM (depending on model)

Transmits 136–174 MHz @ 5 W FM and 420–520 MHz @ 4 W FM, 136–174 MHz @ 5 W FM and 350–470 MHz @ 4 W FM, or 136–174 MHz @ 5 W FM and 216–280 MHz @ 5 W FM (depending on model)

Antenna connector Most ship with SMA F on radio; need SMA M antenna. A few have shipped with the less-common SMA **M** on radio and need an SMA F antenna.

Modes FM

Memory Channels 199

Power No DC input on radio; charger has "M"-style barrel (5.5 mm OD, 2.1 mm ID) with 12 V DC center positive

Model year 2012

Standard Tasks

Program frequency in the field

1. This radio doesn't allow you to overwrite memories. You will need to delete first if you want to write to a memory that has data in it. To delete: press `MENU` `2` `9` `MENU` (DEL-CH) *XXX* where *XXX* is the channel (001–199). Then press `MENU`.

2. If you aren't already in VFO mode, press `MENU` then `S/D`. The screen will show no channel number under the battery icon.

3. Set frequency: use the keypad (144390 for 144.390 MHz).

4. Set transmit tone type and value: press `MENU` `1` `6` `MENU` (T-CTC) then use the knob to select CTCSS tone frequency (or off), press `MENU` `EXIT`. Transmit DCS is set in menu 18, T-DCS.

5. Set receive tone type and value: press `MENU` `1` `5` `MENU` (R-CTC) then use the knob to select CTCSS tone frequency (or off), press `MENU` `EXIT`. Receive DCS is set in menu 17, R-DCS.

6. Set repeater shift: press `MENU` `2` `5` `MENU` (SFT-D) and use the knob to select repeater shift. Press `MENU` `EXIT`.

7. If you need to change the repeater offset frequency: press `MENU` `2` `4` `MENU` (OFFSET) and use the keypad to enter repeater offset (00600 for 600 kHz, 05000 for 5 MHz), then press `MENU` `EXIT`.

8. Set transmit power: press `MENU` `4` `MENU` then use the knob to select power level: LOW (1 W) or HIGH (5 W). Press `MENU` `EXIT`.

9. Write to a memory: press `MENU` `2` `8` `MENU` (`MEM-CH`) and enter channel to write *XXX* (001–199), then press `MENU` `EXIT`.

10. Go to memory mode: press `MENU` `S/D`.

11. Select the memory you just wrote: use the knob.

Lock/unlock radio

Hold `#` for two seconds to lock/unlock.

Check repeater input frequency

Press `*` for less than one second and release. `R` indicator will show you're in reverse mode. Press and release `*` again to return to normal mode.

Change power in the field

To set transmit power, press `MENU` `4` `MENU` then `▲` / `▼` to select power level: `LOW` (1 W) or `HIGH` (5 W). Press `MENU` `EXIT`.

Adjust volume

Rotate the right power/volume knob to adjust volume.

Adjust squelch

Press `MENU` `2` `MENU` (`SQL-LE`) then use the knob to select squelch level (0–9), press `MENU` `EXIT`.

Weird Modes

Radio displays countdown when locking

This is normal for some models; ignore it.

Entering frequency "doesn't take"/can't enter offset

You are probably in channel (memory) mode with frequency display turned on rather than VFO mode. Switch to VFO mode by pressing `MENU` `S/D`.

Useful Information

If you wait too long after pressing `MENU`, the menu mode will time out. You need to press quickly.

There is considerable battery drain when the radio is turned off. Disconnect the battery (or place a piece of paper between the battery terminals and the radio) to preserve the battery when not in use.

This radio has three variants: the KG–UV6D, the KG–UV6D V2 and the KG–UV6X. Front panel programming is identical, but may require different versions of programming software.

Factory reset

To reset the radio, press [MENU] [3] [0] [MENU]. Press [▼] to choose ALL. Press [MENU]. You will see SURE? in the display. (If the device has a password set, you will see - - - - - - and you must enter the correct six-digit password to continue.) Push [MENU] to reset.

Settings reset

To reset the settings, press [MENU] [3] [0] [MENU]. Press [▼] to choose VFO if it isn't already selected. Press [MENU]. You will see SURE? in the display. Push [MENU] to reset.

⚠ **WARNING:** In at least some configurations, the Wouxun KG–UV6D and KG–UV6X may permit you to transmit on business or public safety frequencies. Make sure you are in-band when transmitting.

Radio Layout

Specs

Receivers Two independent receivers, simultaneous receive
Receives 76–108 MHz WFM, 136–174 MHz FM and 420–520 MHz FM, or 136–174 MHz and 400–480 MHz FM (depending on model)
Transmits 136–174 MHz @ 5 W FM and 420–520 MHz @ 4 W FM, or 136–174 MHz @ 5 W FM and 400–480 MHz @ 4 W FM (depending on model)
Antenna connector SMA **M** on radio; needs SMA **F** antenna
Modes FM
Memory Channels 999
Power No DC input on radio; charger has "M"-style barrel (5.5 mm OD, 2.1 mm ID) with 12 V DC center positive
Model year 2014

Standard Tasks

Program frequency in the field

1. If you aren't already in VFO mode, press [VFO/MR]. You may have to press up to three times to cycle through different channel display modes. The screen will show a channel number if you are not in VFO mode.

2. Set frequency: use the keypad (144390 for 144.390 MHz).

3. Set transmit tone type and value: press [MENU] [1] [6] [MENU] (T-CTC) then use the knob to select the CTCSS tone frequency (or OFF). Press [MENU][EXIT]. You can also use R-CTC 15, R-DCS 17 and T-DCS 18 depending on what you want to send/receive.

4. Set repeater shift: press [MENU] [2] [4] [MENU] (SFT-D) and use the knob to select the repeater shift. Press [MENU][EXIT].

5. If you need to change the repeater offset frequency: press [MENU] [2] [3] [MENU] (OFFSET) and enter the repeater offset with the keypad (include leading zeros, 005000 for 5 MHz, 000600 for 600 kHz). Press [MENU][EXIT].

6. Set transmit power: press [MENU] [4] [MENU] and use the knob to select power level: LOW (1 W) or HIGH (5 W VHF, 4 W UHF). Press [MENU][EXIT].

7. Write to a memory: press [MENU] [2] [7] [MENU] (MEM-CH) and enter channel to write on the keypad *XXX* (001–999). Press [MENU][EXIT].

8. Go to memory mode: press [VFO/MR]. You may need to press it up to three times to select the display mode you want.

9. Select the memory you just wrote: use the knob.

Lock/unlock radio

Hold [#] for two seconds to lock/unlock.

Check repeater input frequency

Press ⌜ * ⌟ for less than one second and release. R indicator will show you're in reverse mode. Press and release ⌜ * ⌟ again to return to normal mode.

Change power in the field

To set transmit power, press **MENU** **4** **MENU** and scroll to the power level: LOW (1 W) or HIGH (5 W VHF, 4 W UHF). Press **MENU** **EXIT**.

Adjust volume

Rotate the right power/volume knob to adjust volume.

Adjust squelch

Press **MENU** **2** **MENU** (SQL) then use the knob to select squelch level. Press **MENU** **EXIT**.

Weird Modes

Radio displays circular arrows in upper left

This radio has crossband repeat capability. It is enabled when the circular arrows are displayed in the upper left. Hold **RPT** for two seconds to enable/disable.

You can set the way crossband works with menu 37, RPT-SET. Use X-DIRPT to *receive* on MAIN and transimt on sub-band. Use X-TWRPT for full two-way crossband.

Radio has roger beep

This radio can have a roger beep turned on before, after, or before and after transmission. Use menu 5 ROGER to turn it off.

Entering frequency "doesn't take"

You are probably in channel (memory) mode with frequency display turned on, instead of VFO mode. Switch to VFO mode.

Useful Information

This radio comes in two models, with different ranges for UHF (400–480 or 420–520). The KG-UV8E also includes 220 MHz.

Factory reset

To reset the radio, press **MENU** **5** **1** **MENU**. Press **▼** to choose ALL. Press **MENU**. You will see SURE? in the display. (If the device has a password set, you will see - - - - - - and you must enter the correct six-digit password to continue.) Push **MENU** to reset.

VFO reset

To reset the radio, press `MENU` `5` `1` `MENU`. Press `▼` to choose `VFO` if it isn't already selected. Press `MENU`. You will see `SURE?` in the display. (If the device has a password set, you will see – – – – – – and you must enter the correct six-digit password to continue.) Push `MENU` to reset.

⚠ **WARNING:** In at least some configurations, the Wouxun KG–UV8D may permit you to transmit on business or public safety frequencies. Make sure you are in-band when transmitting.

Wouxun KG-UV9D

Radio Layout

Knob

Power/
Volume

PTT

TDR/VM

MENU

EXIT

*

▼

\#

Plus version pictured

Specs

Receivers Two independent receivers, simultaneous receive
Receives 76–108 MHz WFM (B only), 108–136 MHz AM (A only), 136–174 MHz
    FM, 230–250 MHz FM (A only), 350–400 MHz FM (A only) 400–512 MHz FM,
    700–985 MHz FM (A only)
Transmits 136–174 MHz @ 5 W FM and 400–512 MHz @ 4 W FM
Antenna connector SMA F on radio; needs SMA M antenna
Modes FM
Memory Channels 999
Power No DC input on radio; charger has "M"-style barrel (5.5 mm OD, 2.1 mm ID)
    with 12 V DC center positive
Model year 2015

Standard Tasks

Program frequency in the field

1. If you aren't already in VFO mode, hold **TDR/VM**. You may have to press and release it up to three times to cycle through different channel display modes. The screen will show a channel number if you are not in VFO mode.

2. Set frequency: use the keypad (144390 for 144.390 MHz).

3. Set transmit tone type and value: press **MENU** **1** **7** **MENU** (T-CTC) then use the knob to the CTCSS tone frequency (or OFF). Press **MENU** **EXIT**. You can also use R-CTC 16, R-DCS 18 and T-DCS 19 depending on what you want to send/receive.

4. Set repeater shift: press **MENU** **6** **MENU** (SFT-D). Use the knob to select repeater shift. Press **MENU** **EXIT**.

5. If you need to change the repeater offset frequency: press **MENU** **2** **8** **MENU** (OFFSET) and enter with the keypad (include leading zeros, 005000 for 5 MHz, 000600 for 600 kHz). Press **MENU** **EXIT**.

6. Set transmit power: press **MENU** **5** **MENU** and use the knob to select power level: LOW (1 W), MIDDLE (2 W) or HIGH (5 W VHF, 4 W UHF). Press **MENU** **EXIT**.

7. Write to a memory: press **MENU** **3** **0** **MENU** (MEM-CH) and enter channel to write with the keypad XXX (001–999). Press **MENU** **EXIT**.

8. Go to memory mode: hold **TDR/VM**. You may need to hold it up to three times to select the display mode you want.

9. Select the memory you just wrote: use the knob.

Lock/unlock radio

Hold **#** for two seconds to lock/unlock.

Check repeater input frequency

Press $\boxed{*}$ for less than one second and release. $\mathtt{R}$ indicator will show you're in reverse mode. Press and release $\boxed{*}$ again to return to normal mode.

Change power in the field

To set transmit power, Press $\boxed{\textbf{MENU}}$ $\boxed{\textbf{5}}$ $\boxed{\textbf{MENU}}$ and use the knob to select power level: $\mathtt{LOW}$ (1 W), $\mathtt{MIDDLE}$ (2 W) or $\mathtt{HIGH}$ (5 W VHF, 4 W UHF). Press $\boxed{\textbf{MENU}}$ $\boxed{\textbf{EXIT}}$.

Adjust volume

Rotate the right power/volume knob to adjust volume.

Adjust squelch

Press $\boxed{\textbf{MENU}}$ $\boxed{\textbf{8}}$ $\boxed{\textbf{MENU}}$ ($\mathtt{SQL-LE}$) then use the knob to select squelch level. Press $\boxed{\textbf{MENU}}$ $\boxed{\textbf{EXIT}}$.

Weird Modes

Radio transmits on wrong frequency

The plus version of this radio has crossband repeat capability. Set menu 43, $\mathtt{TYPE-SET}$ to $\mathtt{TALKIE}$ to disable.

Radio has roger beep

This radio can have a roger beep turned on before, after, or before and after transmission. Use menu 9 $\mathtt{ROGER}$ to turn it off.

Entering frequency "doesn't take"

You are probably in channel (memory) mode with frequency display turned on, instead of VFO mode. Switch to VFO mode.

Useful Information

There are two models, the KG–UV9D and the KG–UV9D+. The plus version has 61 menu entries rather than the 55 of the non–plus version, and includes crossband repeat.

Factory reset

To reset the radio, press $\boxed{\textbf{MENU}}$ $\boxed{\textbf{6}}$ $\boxed{\textbf{1}}$ $\boxed{\textbf{MENU}}$ (plus version) or $\boxed{\textbf{MENU}}$ $\boxed{\textbf{5}}$ $\boxed{\textbf{5}}$ $\boxed{\textbf{MENU}}$ (non–plus version). Press $\boxed{\blacktriangledown}$ to choose $\mathtt{ALL}$. Press $\boxed{\textbf{MENU}}$. You will see $\mathtt{SURE?}$ in the display. (If the device has a password set, you will see $\mathtt{-\ -\ -\ -\ -\ -}$ and you must enter the correct six–digit password to continue.) Push $\boxed{\textbf{MENU}}$ to reset.

VFO reset

To reset the radio, press [MENU] [6] [1] [MENU] (plus version) or [MENU] [5] [5] [MENU] (non-plus version). Press [▼] to choose VFO if it isn't already selected. Press [MENU]. You will see SURE? in the display. (If the device has a password set, you will see ‑ ‑ ‑ ‑ ‑ ‑ and you must enter the correct six-digit password to continue.) Push [MENU] to reset.

⚠ **WARNING:** In at least some configurations, the Wouxun KG–UV9D may permit you to transmit on business or public safety frequencies. Make sure you are in-band when transmitting.

Radio Layout

Specs

Receivers Single receiver with priority channel option
Receives 110–138 MHz AM, 136–180 MHz FM
Transmits 144–148 MHz @ 5 W FM
Antenna connector BNC F on radio; needs BNC M antenna
Modes FM
Memory Channels 146 (71 if using alpha tags)
Power No DC input on radio; 4–12 V DC per spec
Model year 1993

Standard Tasks

Program frequency in the field

1. If you aren't already in VFO mode, press **VFO**. The screen will show a channel number in the upper left when not in VFO mode.

2. Set frequency: use the keypad, omitting first two digits (enter 4390 for 144.390 MHz).

3. Set transmit tone type: press **F/M** **1** (TONE). The display will show **T** in upper part of screen when tone is on. Press **F/M** again to leave function mode. If tone decode option is installed, radio will cycle through **T** (send tone on transmit), **T SQ** (send tone on transmit, require tone on receive to break squelch) and none.

4. Set repeater shift: press **F/M** **6** (RPT) to cycle through **+**, **-** and nothing (simplex). Press **F/M** again to leave function mode.

5. Set transmit tone: press **F/M** **2** (T SET). Use knob or **▲**/**▼** to set tone desired CTCSS tone. Press **2** when done.

6. Set transmit power: press **F/M** **3** LOW to set power level as described in the section "Change power in the field," below.

7. Write to a memory: hold **F/M** for at least half a second.

8. Select the desired memory using the knob or **▲**/**▼**.

9. Press and release **F/M** to write.

10. Go to memory mode: press **MR**.

11. Select the memory you just wrote: use the knob.

12. For odd splits, program the receive frequency as described above. Then press **VFO** to go back to VFO mode, enter transmit frequency on keypad omitting first two digits (7840 for 147.840). Hold **F/M** for half a second. Use the knob to select the memory that holds RX frequency. Hold **PTT** (does not transmit) and press **F/M** to write transmit frequency. Successfully written memory will show **+-** to indicate odd split.

Lock/unlock radio

Flip lock switch up to lock (screen shows KL for keyboard lock, VL for volume lock, DL for knob lock and PL for PTT lock). Flip down to unlock.

Check repeater input frequency

Press F/M 9 (REV) to switch between reverse and regular modes.

Change power in the field

1. To set transmit power, press and release F/M 3 (LOW). You'll see either LOW or HIGH. Press 3 to change between them. LOW ranges from 0.3–3 W, and HIGH ranges from 1.5–5 W, depending on the battery installed.

2. On LOW, press ▲ ▼ to choose between LOW 1 (0.3 W), LOW 2 (1.5 W) and LOW 3 (1.5 or 3 W, depending on battery installed).

3. Wait three seconds or press and release PTT (does not transmit) to set.

Adjust volume

Press VOL ▲ or VOL ▼ to adjust volume.

Adjust squelch

Press F/M. Within three seconds, press VOL ▲ or VOL ▼ to adjust squelch.

Weird Modes

Radio blinks LOW

Battery overheat temperature sensor has triggered. The radio automatically switches to low power when that happens. Transmit less and wait for the battery to cool down.

Radio doesn't receive when someone else transmits

This radio has a DTMF squelch option. Press PAGE repeatedly until you don't see PAGE, T.PAGE or CODE in the display.

Knob changes volume and not frequency

Press and hold both VOL ▲ and VOL ▼ when powering on.

Entering frequency as four digits doesn't work

The radio may be in extended frequency range mode (see the "Useful Information" section). Enter the frequency as six digits: 144390 for 144.390 MHz.

ERR when PTT pressed

This radio displays ERR when you try to transmit out of band. Check frequency, repeater shift and repeater offset.

Can't enter VFO mode

This radio has a memory-only mode. To enable or disable it, turn the radio off. Then hold [F/M] and [MR] while turning the radio on.

Useful Information

Hold [Power] for half a second to turn the radio on or off.

The radio can transmit CTCSS tones as shipped, but must have the FTS-26 option installed to do tone decoding/tone squelch.

Default offset can be changed. Press [F/M] [0] (SET) to enter set mode. Use knob to select menu item 6 (usually 0.600). Use [▲] [▼] to change. Press [PTT] (does not transmit) to save.

Lock type can be changed. Press [F/M] [0] (SET) to enter set mode. Use knob to select menu item 5 (LOCK). Use [▲] to change PTT lock (PL) and keyboard lock (KL). Use [▼] to change volume lock (VL) and knob lock (DL). Press [PTT] (does not transmit) to save.

Extended frequency range mode: to set RX frequencies to extended range (110–180 MHz), press and hold both [▲] and [▼] while turning radio on. Frequencies below 136 MHz will be AM.

Factory reset

To reset the radio, hold [MR], [VFO] and [2] while turning the radio on. This will reset all settings and memories.

Radio Layout

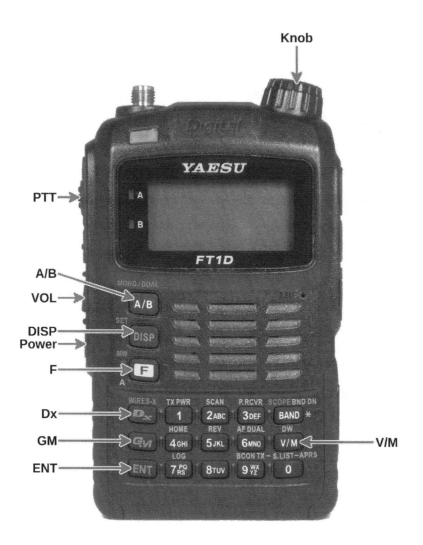

Knob

PTT

A/B

VOL

DISP
Power

F

Dx

GM

ENT

V/M

Specs

Receivers Two independent receivers, simultaneous receive

Receives 500 kHz–76 MHz AM on VFO A, 76 MHz–108 MHz FM VFO A, 108 MHz–137 MHz AM on VFO A, 137 MHz–774 MHz FM on VFO A, 803 MHz–999 MHz FM on VFO A (cell blocked)

Transmits 144–148 MHz @ 5 W FM, 430–450 MHz @ 5 W FM

Antenna connector SMA F on radio; needs SMA M antenna

Modes FM

Memory Channels 900

Power 4–14 V DC, EIAJ-02 barrel style, 4mm OD, 1.7mm ID plug, center positive

Model year 2012

Standard Tasks

Program frequency in the field

1. If you aren't already in VFO mode, press **V/M**. The screen will show **VFO**.

2. Set frequency: use the keypad (144390 for 144.390 MHz).

3. Set transmit tone type: hold **DISP** for at least one second. Use the knob to select **4 SIGNALING**. Press **ENT**. Use the knob to select **11 SQL TYPE**. Press **ENT**. Use the knob to select from **OFF** (no tone), **TONE TN** (CTCSS tone on transmit), **TONE SQL TSQ** (CTCSS tone on transmit and required for receive), **DCS DCS** (DCS), **REV TONE RTN** (squelch if tone received), **PR FREQ PR** ("no-communication squelch" with a frequency) and **PAGER PAG** (detect two CTCSS tones). Press **DISP** repeatedly to exit the menu system and set the value.

4. Set transmit tone: hold **DISP** for at least one second. Use the knob to select **4 SIGNALING**. Press **ENT**. Use the knob to select **12 TONE SQL FREQ**. Press **ENT**. Use the knob to select the tone. Press **DISP** repeatedly to exit the menu system and set the value. Use menu **2 DCS CODE** to set the DCS code.

5. Set repeater shift: hold **DISP** for at least one second. Use the knob to select **4 SIGNALING**. Press **ENT**. Use the knob to select **14 RPT SHIFT**. Press **ENT**. Use the knob to select from **+RPT** (positive shift), **-RPT** (negative shift) and **SIMPLEX** (no shift). Press **DISP** repeatedly to exit the menu system and set the value. Note that this radio has automatic repeater shift, but it can be overridden.

6. If you need to change the repeater offset frequency: hold **DISP** for at least one second. Use the knob to select **4 SIGNALING**. Press **ENT**. Use the knob to select **16 RPT SHIFT FREQ**. Press **ENT**. Use the knob to adjust the offset. Press **DISP** repeatedly to exit the menu system and set the value.

7. Set transmit power: press **F** **1**. Use the knob to cycle through **HIGH** (5 W), **LOW3** (2.5 W), **LOW2** (1 W) and **LOW1** (0.1 W). Press **F** to save.

8. Write to a memory: hold [F] for at least one second. Use the knob to select the memory to write. The upper-left shows a page with lines on it if the memory already has data, and an empty page if the memory has not yet been programmed. Press [F] to program.

9. Go to memory mode: press [V/M].

10. Select the memory you just wrote: use the knob.

Lock/unlock radio

Press and release [Power] for less than one second to lock/unlock.

Check repeater input frequency

Press [F] then [5] switch between reverse and regular modes. The shift indicator will flash when you are in reverse.

Change power in the field

To set transmit power, press [F][1]. Use the knob to cycle through HIGH (5 W), LOW3 (2.5 W), LOW2 (1 W) and LOW1 (0.1 W). Press [F] to save.

Adjust volume

Hold [VOL] and rotate the knob to adjust. Each VFO has its own volume.

Adjust squelch

Hold [DISP] for at least one second. Use the knob to select 4 SIGNALING. Press [ENT]. Use the knob to select 8 SQL LEVEL. Press [ENT]. Use the knob to adjust the squelch from LEVEL0 (open) through LEVEL15. Press [DISP] repeatedly to exit the menu system and set the value.

Weird Modes

Radio shows DN or VW instead of FM

You are in a digital mode. Press the [Dx] button to cycle through the modes. A mode with a black box in front indicates you are in automatic mode select, where the radio will pick the mode based on what it receives.

Useful Information

Hold [Power] for more than one second to turn the radio on or off.

Hold [A/B] for more than one second to switch between monitoring two frequencies and monitoring one frequency. Press and release [A/B] to switch between bands.

This radio has multiple lock modes. To change the way the radio locks, hold [DISP] for at least one second. Use the knob to select 8 CONFIG. Press [ENT]. Use the knob to select 9 LOCK. Press [ENT]. Use the knob to cycle through KEY&DIAL (keyboard and dial locked out), PTT (only [PTT] locked out), KEY&PTT (keyboard and PTT locked out), DIAL&PTT (dial and [PTT] locked out but keyboard active), ALL (dial, keyboard and [PTT] locked out), KEY (only keyboard locked out) and DIAL (only dial locked out).

The first version of this radio, the FT-1D, was available in black or silver. The later version, the FT-1XD, came out in 2015 and was available in black only. The FT-1XD has an improved GPS but is otherwise the same radio as the FT-1D.

There are two generations of the FT-1D. GEN1 can receive 222–420 MHz but not 803–999 MHz, while GEN2 can receive 803–999 MHz (cell blocked) but not 222–420 MHz.

Factory reset

To reset everything, hold [Dx], [GM] and [ENT] while turning the radio on. The radio will display ALL RESET PUSH F KEY! and at this point you will reset if you press [F]. After a reset, you will be prompted to input a callsign, which you can do using the keypad (press multiple times to cycle through each letter). Press [PTT] (does not transmit) to save the callsign and start using the radio.

Settings reset

To reset most settings but leave memories intact, hold [Dx] and [V/M] while turning the radio on, then press [F].

Yaesu FT-209R/FT-209RH

Radio Layout

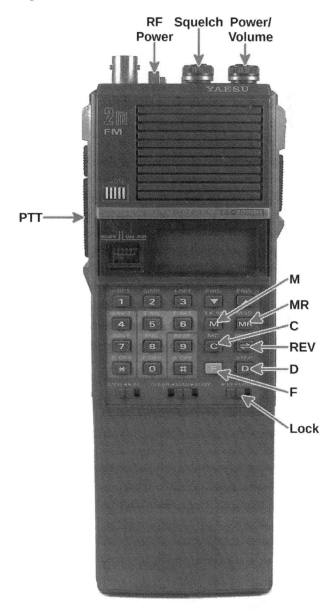

RF Power   Squelch   Power/Volume

PTT

M
MR
C
REV
D
F
Lock

Specs

Receivers Single receiver
Receives 144–148 MHz FM
Transmits 144–148 MHz @ 3.5 W FM (209R), 144–148 MHz @ 5 W FM (209RH)
Antenna connector BNC F on radio; needs BNC M antenna
Modes FM
Memory Channels 10
Power 6–15 V DC in spec, nominally 12 V DC on radio. Power adapter on FNB-4
    battery: IEC 60130-10 barrel style type A, 5.5mm OD, 2.5mm ID plug, **center
    negative**. Charging plug on FNB-4 battery: 2.5mm tip/sleeve phone plug, tip
    positive.
Model year 1985

Standard Tasks

Program frequency in the field

1. Decide which memory you want to write to (0–9). Note that 0 is special (can't
   be deleted, and is used for calling channel). In these instructions, n represents
   the number key on the keyboard which corresponds to that memory.

2. Set transmit frequency: enter the last four digits of frequency (so 4390 is 144.390
   MHz). On B and C versions, the last digit is assumed (enter 438 for 144.3875
   MHz). Press D to set the frequency. Press n and then M to write.

3. Set repeater shift: press F then press 1 (-RPT) or 3 (+RPT) for negative or
   positive shift. To set non-standard shift, instead enter the transmit frequency you
   want. Then press D, n F and then M (TXM). Press n and then M to
   write.

4. Set transmit tone: pick the correct one or two digit value from the table below,
   enter it, then press F and then 6 (T SET). Press n and then M to write.

5. Set transmit tone type: press F and then 8 (ENC). (For tone squelch, press
   F and then 5 (T SQ) instead.) Then press n and then M to write.

6. Select the memory you just wrote: press n and then MR

Lock/unlock radio

Slide the keylock switch on the radio front panel to the right to lock. Slide to the left
to unlock.

Check repeater input frequency

Press and release REV to switch between reverse and regular modes

Change power in the field

To set transmit power, press the RF pushbutton switch on top of the radio. Press it
down low power (350 mW on FT-209R, 500 mW on FT-209RH). Press it up for high
power (3.5 W on FT-209R, 5 W on FT-209RH).

Adjust volume

Rotate the right power/volume knob to adjust volume.

Adjust squelch

Rotate the left squelch knob to adjust RF squelch.

Weird Modes

Can't enter some frequencies

The radio comes in four versions (A, B, C and E). The A and E versions have 5 kHz channel spacing. The B and C versions have 12.5 kHz channel spacing, and will not let you enter any frequency which has a 4 or 9 in the 10 kHz digit, since those don't correspond to 12.5 kHz channel spacing. There is no way to enter those frequencies on the B and C versions.

Radio transmits unexpectedly

If you're using a headset, be sure you don't have VOX turned on. Press the VOX ON switch until it is up (off). VOX will not work without a headset.

1000 Hz tone when transmitting

In addition to the regular CTCSS tones, the FT-209 includes a test tone of 1000 Hz. Switch to a different tone (or none at all) to get rid of it.

Useful Information

The B version has a frequency range of 144–146 MHz rather than 144–148 MHz. The radio requires a FTS-6 tone board to generate CTCSS tones. Only the A version of the radio supports this board. (The A version can be recognized because it does not have a BURST above the PTT.)

DTMF is supported only on the A version of the receiver.

There is a switch on the radio underneath the battery pack that doubles channel spacing (goes from 5 to 10 kHz on A and E, 12.5 to 25 on B and C). You must remove the battery pack to see this switch, and will need a paperclip or other small object to press it.

The switches on the front panel are: S/PO: meter shows signal strength on receive, power on transmit. BC: meter shows battery. Clear: scan stops on empty channel. Man: scan stops when scan key is released. Busy: scan stops on busy channel.

To delete any memory except 0: press n then press MR, then press F , C , D.

To see which memories are in use, press D , M , M.

CTCSS tones (Hz) and their FT-209 code values

67.0: 34	91.5: 42	123.0: 13	162.2: 21	218.1: 29
71.9: 35	94.8: 6	127.3: 14	167.9: 22	225.7: 30
74.4: 36	100.0: 7	131.8: 15	173.8: 23	233.6: 31
77.0: 37	103.5: 8	136.5: 16	179.9: 24	241.8: 32
79.7: 38	107.2: 9	141.3: 17	186.2: 25	250.3: 55
82.5: 39	110.9: 10	146.2: 18	192.8: 26	
85.4: 40	114.8: 11	151.4: 19	203.5: 27	
88.5: 41	118.8: 121	156.7: 20	210.7: 28	1000: 63

Factory reset

To reset, insert a paperclip in the hole on the back of the case. This resets all memories and settings.

Radio Layout

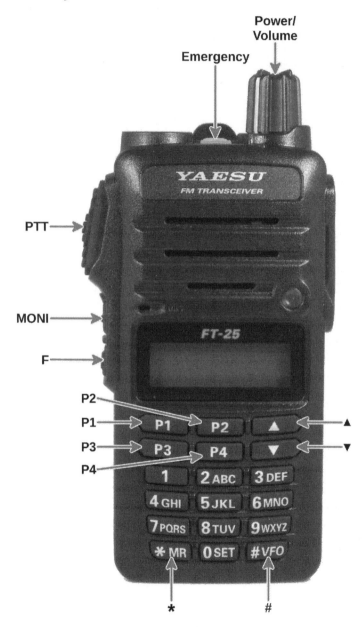

Specs

Receivers Single receiver with priority channel option
Receives 65–108 MHz WFM, 136–174 MHz FM
Transmits 144–148 MHz @ 5 W FM
Antenna connector SMA **M** on radio; needs SMA **F** antenna
Modes FM
Memory Channels 200
Power No power input on radio, nominally 7.4 V DC
Model year 2017

Standard Tasks

Program frequency in the field

1. If you aren't already in VFO mode, press `#`. The screen shows `M` if you're in memory mode.

2. Set frequency: use the keypad (144390 for 144.390 MHz).

3. Hold `F` to enter menu mode.

4. Set transmit tone: use `▲`/`▼` to go to item 8, `CTCSS`. Press `F` to select `TX` if desired, then `▲`/`▼` to adjust. Press `F` to set. Similarly, use `▲`/`▼` to select `RX` if desired, then `▲`/`▼` to adjust. Press `F` to set.

5. Set repeater shift: use `▲`/`▼` to go to item 24, `Repeater`. Press `F` to select `Mode`. Use `▲`/`▼` to switch between `Simplex`, `+REP` (positive offset) and `-REP` (negative offset). Press `F` to set.

6. If you need to change the repeater offset frequency: use `▼` to go to item 25, `Shift`. Press `F` to change, and use `▲`/`▼` to adjust. Press `F` to set when done.

7. Set transmit tone type: use `▲`/`▼` to go to item 29, `SQL Type`. Press `F` to adjust. Use `▲`/`▼` to select from `R-TONE` (receive squelch), `T-TONE` (transmit tone), `TSQL` (transmit tone and receive tone squelch), `REV TN` (squelch if tone received), `DCS` (digital coded squelch), `PAGER` (unsquelch if two CTCSS tones received) , and `OFF` (no tone). Press `F` to set.

8. Set transmit power: use `▲`/`▼` to go to item 32, `TX PWR`. Press `F` to adjust. Use `▲`/`▼` to choose from `HI(5 W)`, `MID(2.5 W)` or `LOW(0.5 W)`. Press `F` to set.

9. Hold `F` to exit the menu.

10. Write to a memory: hold `*`.

11. Use `▲`/`▼` to select the memory.

12. Hold `*` to complete writing to memory.

13. Go to memory mode: press [*].

14. Select the memory you just wrote: use [▲][▼].

Lock/unlock radio

Hold [6] to lock/unlock.

Check repeater input frequency

Press [F][P4] to enter or exit reverse mode. Shift will blink when you're in reverse mode.

Change power in the field

To set transmit power, hold [F] to enter the menu. Use [▲][▼] to go to item 32, TX PWR. Press [F] to adjust. Use [▲][▼] to choose from HI(5 W), MID(2.5 W) or LOW(0.5 W). Press [F] to set. Hold [F] to exit the menu.

Adjust volume

Rotate the knob to adjust volume.

Adjust squelch

Hold [F] to enter hte menu. Scroll to item 26, RF SQL. Press [F] to adjust. Use [▲][▼] to go from OFF, S-1 through S-8, or S-FULL. Press [F] to set. Hold [F] to exit the menu.

Weird Modes

Radio is beeping continually

You are in emergency mode. Turn the radio off and then on. If you hold down [Emergency] for three seconds, you will enter emergency mode. This causes a continual beep and wil transmit on the home frequency.

Can't enter VFO mode

This radio has a memory-only mode. To exit, turn the radio off. Then hold [MONI] and [PTT] (does not transmit) while turning the radio on. Use [▼] to select F5:MEM-ONLY. Press [F] to toggle.

Can't enter VHF frequencies

This radio has an option to use FM only. To adjust it, turn the radio off. Then hold [MONI] and [PTT] (does not transmit) while turning the radio on. Use [▼] to select F7:FM-ONLY. Press [F] to toggle.

Radio shows OUT.RNG or IN.RNG

This means you have ARTS mode enabled. Hold [2] to exit.

Radio shows - - - - on power on

The password has been enabled. Enter the password (four digits) on the keypad. If you have forgotten the password, you can disable with a full reset (see below).

Useful Information

This radio has four programmable buttons which remember the radio state. Rather than programming a memory, once you've set up the frequency, hold [P1], [P2], [P3] or [P4] to program the button. Then later you can press [P1], [P2], [P3] or [P4] to recall the state (including frequency/tone info).

To enable expanded transmit range (137–174 MHz) turn the radio off. Then hold down [MONI] and [PTT] (does not transmit) while turning the radio on. You will see F1 SET RESET. Using the keypad, enter 32406665 (some versions require 22406665 or 62406665 instead). The radio will reset. At this point you can transmit out of band. Repeat the procedure to go back to the default.

Factory reset

To reset, hold [MONI] and [PTT] (does not transmit) while turning the radio on. Use [▲]/[▼] to select F4 ALL RESET (full reset—reset everything). Press [F] to reset.

Memory reset

To reset, hold [MONI] and [PTT] (does not transmit) while turning the radio on. Use [▲]/[▼] to select F2 MEM RESET (reset memories). Press [F] to reset.

Memory bank reset

To reset, hold [MONI] and [PTT] (does not transmit) while turning the radio on. Use [▲]/[▼] to select F3 BANK RESET (reset memory banks). Press [F] to reset.

Settings reset

To reset, hold [MONI] and [PTT] (does not transmit) while turning the radio on. Use [▲]/[▼] to select F1 SET RESET (reset settings). Press [F] to reset.

Radio Layout

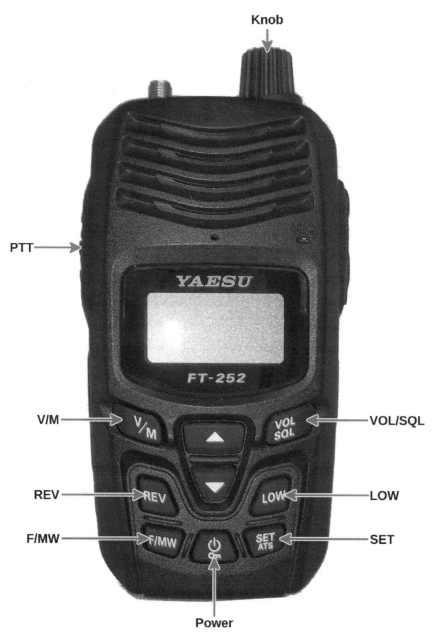

Knob

PTT

V/M

REV

F/MW

VOL/SQL

LOW

SET

Power

Specs

Receivers Single receiver
Receives 136–174 MHz FM
Transmits 144–148 MHz @ 5 W FM
Antenna connector SMA F on radio; needs SMA M antenna
Modes FM
Memory Channels 200
Power Nominally 5–10 V DC, but charger is 10.5 V DC, EIAJ-02 barrel style, 4mm
    OD, 1.7mm ID plug, center positive
Model year 2013

Standard Tasks

Program frequency in the field

1. If you aren't already in VFO mode, press **V/M**. The screen will show no channel number in VFO mode.

2. Set frequency: use the knob. Press **F/MW** to adjust in 1 MHz steps.

3. Enter menu mode: press **SET**.

4. Set transmit tone: use the knob to select menu `42:TN FRQ`. Press **SET** to adjust tone frequency. Adjust with the knob, and press **SET** when done. Use `13:DCS.COD` for DCS tones instead.

5. Set transmit tone type: use the knob to select menu `40:SQL.TYP`. Press **SET** and use the knob to select `TONE` (CTCSS tone on transmit), `TSQL` (tone on transmit and receive), `REV TN` (squelch on CTCSS tone receive), `DCS` (DCS), `D` (DCS encode only), `T DCS` (CTCSS transmit, DCS for receive), `D TSQL` (DCS transmit, CTCSS receive) or `OFF`. Press **SET** when done.

6. Set repeater shift: use the knob to select menu `32:RPT.MOD`. Press **SET** to adjust. Use the knob to select from `RPT. +`, `RPT.OFF` (simplex), or `RPT. -`. Press **SET** when done.

7. If you need to change the repeater offset frequency: use the knob to select menu `28:R SHFT`. Press **SET** then use knob to adjust repeater offset. Press **SET** when done.

8. Exit menu mode: press **F/MW**.

9. Set transmit power: press **LOW** repeatedly. Options are `HIGH` (5 W)/`MID` (2 W)/`LOW` (0.5 W). Wait about two seconds for the mode to be set.

10. Write to a memory: hold **F/MW** for at least half a second.

11. Use the knob to select the desired memory.

12. Press and release **F/MW** to write.

Lock/unlock radio

Press **Power** to lock/unlock.

Check repeater input frequency

Press and release **REV** to switch between reverse and regular modes. For this to work, the radio must have menu `30:REV/HM` set to `REV` mode.

Change power in the field

To set transmit power, press **LOW** repeatedly. Options are `HIGH` (5 W)/`MID` (2 W)/`LOW` (0.5 W). Wait about two seconds for the mode to be set.

Adjust volume

Press **VOL/SQL** until you see `VOL`, then rotate knob to adjust volume. The radio automatically leaves this mode after about three seconds with no action.

Adjust squelch

Press **VOL/SQL** until you see `SQL`, then rotate knob to adjust volume. The radio automatically leaves this mode after about three seconds with no action.

Weird Modes

Radio is beeping continually

You are in emergency mode. Turn the radio off and then on.

Radio shows `OUT.RNG`, `IN.RNG` or `SYNC`

This means you have ATS mode enabled. Press **F/MW** and then **SET** to disable.

Radio shows - - - - on power on

The password has been enabled. Use the knob to select and press **F/MW** to enter each character at a time. If you have forgotten the password, you can disable with a full reset (see below).

Can't get to VFO mode

The radio has a memory-only mode. To disable, hold **V/M** while turning the radio on, then select `F5:M-ONLY`, then press **SET**. This toggles from memory-only mode to regular mode.

Useful Information

Hold [**Power**] for about two seconds to turn off or on.

If you hold [**SET**] for one second, you will enter emergency mode. This causes a continual beep and can transmit if configured to do so using menu `19:EMG S`. Options are `EMG.BEP` (beep), `EMG.LMP` (flash the display), `EMG.B+L` (beep + display), `EMG.CWT` (transmit SOS), `EMG.C+B` (transmit + beep), `EMG.C+L` (transmit + display), `EMG.ALL` (transmit + beep + display) and (mercifully) `OFF`.

In order to see some transmit tone options, menu `39:SPLIT` must be set to `SPL ON`.

This radio has no way to program it other than to do so by the front panel. Be cautious about resetting memories, since the radio user will have to manually re-enter everything.

Factory reset

To reset, hold [**V/M**] and turn the radio on. Use the knob to select `F4 ALLRST` (full reset—reset everything). Press [**SET**] to reset.

Memory reset

To reset, hold [**V/M**] and turn the radio on. Use the knob to select `F2 MEMRST` (reset memories). Press [**SET**] to reset.

Memory bank reset

To reset, hold [**V/M**] and turn the radio on. Use the knob to select `F3 MB RST` (reset memory banks). Press [**SET**] to reset.

Settings reset

To reset, hold [**V/M**] and turn the radio on. Use the knob to select `F1 SETRST` (reset settings). Press [**SET**] to reset.

Yaesu FT-257

Radio Layout

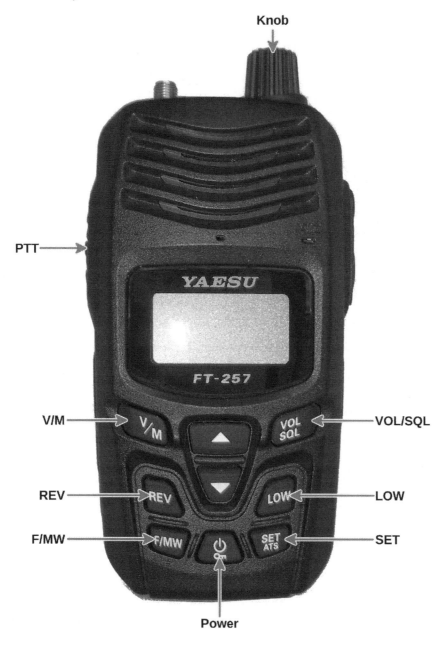

Knob

PTT

V/M

REV

F/MW

Power

VOL/SQL

LOW

SET

Specs

Receivers Single receiver
Receives 400–480 MHz FM
Transmits 430–450 MHz @ 5 W FM
Antenna connector SMA F on radio; needs SMA M antenna
Modes FM
Memory Channels 200
Power Nominally 5–10 V DC, but charger is 10.5 V DC, EIAJ-02 barrel style, 4mm
    OD, 1.7mm ID plug, center positive
Model year 2013

Standard Tasks

Program frequency in the field

1. If you aren't already in VFO mode, press **V/M**. The screen will show no channel number in VFO mode.

2. Set frequency: use the knob. Press **F/MW** to adjust in 1 MHz steps.

3. Enter menu mode: press **SET**.

4. Set transmit tone: use the knob to select menu `42:TN FRQ`. Press **SET** to adjust tone frequency. Adjust with the knob, and press **SET** when done. Use `13:DCS.COD` for DCS tones instead.

5. Set transmit tone type: use the knob to select menu `40:SQL.TYP`. Press **SET** and use the knob to select `TONE` (CTCSS tone on transmit), `TSQL` (tone on transmit and receive), `REV TN` (squelch on CTCSS tone receive), `DCS` (DCS), `D` (DCS encode only), `T DCS` (CTCSS transmit, DCS for receive), `D TSQL` (DCS transmit, CTCSS receive) or `OFF`. Press **SET** when done.

6. Set repeater shift: use the knob to select menu `32:RPT.MOD`. Press **SET** to adjust. Use the knob to select from `RPT. +`, `RPT.OFF` (simplex), or `RPT. -`. Press **SET** when done.

7. If you need to change the repeater offset frequency: use the knob to select menu `28:R SHFT`. Press **SET** then use knob to adjust repeater offset. Press **SET** when done.

8. Exit menu mode: press **F/MW**.

9. Set transmit power: press **LOW** repeatedly. Options are `HIGH` (5 W)/`MID` (2 W)/`LOW` (0.5 W). Wait about two seconds for the mode to be set.

10. Write to a memory: hold **F/MW** for at least half a second.

11. Use the knob to select the desired memory.

12. Press and release **F/MW** to write.

Lock/unlock radio

Press [**Power**] to lock/unlock.

Check repeater input frequency

Press and release [**REV**] to switch between reverse and regular modes. For this to work, the radio must have menu `30:REV/HM` set to `REV` mode.

Change power in the field

To set transmit power, press [**LOW**] repeatedly. Options are `HIGH` (5 W)/`MID` (2 W)/`LOW` (0.5 W). Wait about two seconds for the mode to be set.

Adjust volume

Press [**VOL/SQL**] until you see `VOL`, then rotate knob to adjust volume. The radio automatically leaves this mode after about three seconds with no action.

Adjust squelch

Press [**VOL/SQL**] until you see `SQL`, then rotate knob to adjust volume. The radio automatically leaves this mode after about three seconds with no action.

Weird Modes

Radio is beeping continually

You are in emergency mode. Turn the radio off and then on.

Radio shows `OUT.RNG`, `IN.RNG` or `SYNC`

This means you have ATS mode enabled. Press [**F/MW**] and then [**SET**] to disable.

Radio shows - - - - on power on

The password has been enabled. Use the knob to select and press [**F/MW**] to enter each character at a time. If you have forgotten the password, you can disable with a full reset (see below).

Can't get to VFO mode

The radio has a memory-only mode. To disable, hold [**V/M**] while turning the radio on, then select `F5:M-ONLY`, then press [**SET**]. This toggles from memory-only mode to regular mode.

Useful Information

Hold [Power] for about two seconds to turn off or on.

If you hold [SET] for one second, you will enter emergency mode. This causes a continual beep and can transmit if configured to do so using menu `19:EMG S`. Options are `EMG.BEP` (beep), `EMG.LMP` (flash the display), `EMG.B+L` (beep + display), `EMG.CWT` (transmit SOS), `EMG.C+B` (transmit + beep), `EMG.C+L` (transmit + display), `EMG.ALL` (transmit + beep + display) and (mercifully) `OFF`.

In order to see some transmit tone options, menu `39:SPLIT` must be set to `SPL ON`.

This radio has no way to program it other than to do so by the front panel. Be cautious about resetting memories, since the radio user will have to manually re-enter everything.

Factory reset

To reset, hold [V/M] and turn the radio on. Use the knob to select `F4 ALLRST` (full reset—reset everything). Press [SET] to reset.

Memory reset

To reset, hold [V/M] and turn the radio on. Use the knob to select `F2 MEMRST` (reset memories). Press [SET] to reset.

Memory bank reset

To reset, hold [V/M] and turn the radio on. Use the knob to select `F3 MB RST` (reset memory banks). Press [SET] to reset.

Settings reset

To reset, hold [V/M] and turn the radio on. Use the knob to select `F1 SETRST` (reset settings). Press [SET] to reset.

Yaesu FT-270

Radio Layout

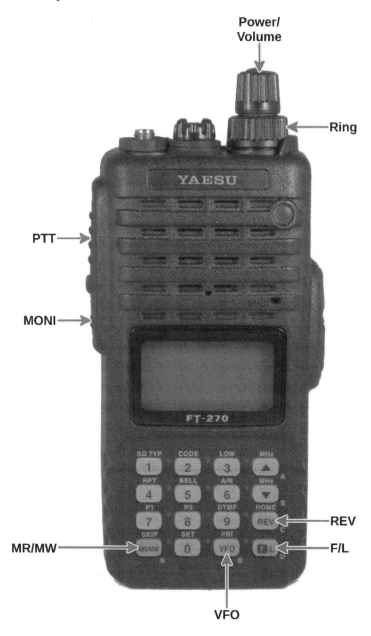

Power/Volume

Ring

PTT

MONI

REV

F/L

MR/MW

VFO

Specs

Receivers Single receiver, option of priority channel or dual watch (first to break squelch wins)
Receives 136–174 MHz FM
Transmits 144–148 MHz @ 5 W FM
Antenna connector SMA F on radio; needs SMA M antenna
Modes FM
Memory Channels 200
Power 6.0–16.0 V DC, EIAJ-02 barrel style, 4mm OD, 1.7mm ID plug, center positive
Model year 2009

Standard Tasks

Program frequency in the field

1. If you aren't already in VFO mode, press **VFO**. The display will show `-A-` or `-b-` in VFO mode, and a memory number in memory mode.

2. Set frequency: use the keypad (144390 for 144.390 MHz).

3. Set transmit tone type: press **F/L** **1** (SQ TYPE). Use the ring to adjust. Values are `OFF` (no tone), `TONE` (tone on transmit), `TSQL` (tone sent on transmit and required for receive), `REV TN` (tone causes radio not to receive), `DCS` (digital coded squelch), `ECS` (enhanced paging squelch). Press **PTT** (does not transmit) to set.

4. Set repeater shift: press **F/L** **4** (RPT). Use the ring to adjust. Values are `RPT. +` (positive shift), `RPT.OFF` (simplex) and `RPT. -` (negative shift). Press **F/L** to set.

5. Set transmit tone: press **F/L** **0** (SET) to enter the set menu. Use the ring to select item 46, `TN FRQ`. Press **F/L** to change. Use the ring to adjust, and press **F/L** to set.

6. If you need to change the repeater offset frequency: press **F/L** **0** (SET) to enter the set menu. Use the ring to select item 41, `SHIFT`. Press **F/L** to change. Use the ring to adjust, and press **F/L** to set.

7. Press **PTT** (does not transmit) to leave menu mode.

8. Set transmit power: press **F/L** **3** (LOW). Use the ring to select the level you want: (`HIGH` (5 W), `MID` (2 W), or `LOW` (0.5 W). Press and release **F/L** to set the power level.

9. Write to a memory: hold **MR/MW** for at least one second.

10. Use the ring to select the desired memory to write.

11. Press **MR/MW** to write.

12. Go to memory mode: press **MR/MW**.

13. Select the memory you just wrote: use the ring.

Lock/unlock radio

Hold [F/L] for one second to lock/unlock.

Check repeater input frequency

Press and release [REV] to switch transmit and receive frequencies. The repeater offset will blink when you are reversed.

Change power in the field

To set transmit power, press [F/L] [3] (LOW). Use the ring to select the level you want: (HIGH (5 W), MID (2 W), or LOW (0.5 W). Press and release [F/L] to set the power level.

Adjust volume

Rotate the center power/volume knob to adjust volume.

Adjust squelch

Press [F/L] [MONI]. Rotate the ring to adjust squelch (1–15). Press [PTT] (does not transmit) to save.

Weird Modes

Radio shows OUT.RNG

This means you have ARTS mode enabled. Press [F/L] to exit ARTS mode.

Can't enter VFO mode

This radio has a memory-only mode. To enable/disable it, hold [MONI] while turning the radio on. Use the ring to select F5 M-ONLY and press [F/L].

Radio doesn't hear transmissions

This radio has a feature to prevent signals from opening squelch unless they have a certain S-meter reading. To change this, press [F/L] [0] (SET). Rotate the ring to choose menu 34, RF SQL. Press [F/L] to adjust. Rotate the ring to set to OFF. Press [PTT] (does not transmit) to save.

Useful Information

Factory reset

To reset the radio, hold [MONI] while turning the radio on. Use the ring to select F4 ALLRST (reset everything). Press [F/L] to reset.

Memory reset

To reset the radio, hold [**MONI**] while turning the radio on. Use the ring to select F2 MEMRST (clear memories). Press [**F/L**] to reset.

Memory bank reset

To reset the radio, hold [**MONI**] while turning the radio on. Use the ring to select F3 MB RST (clear memory banks). Press [**F/L**] to reset.

Settings reset

To reset the radio, hold [**MONI**] while turning the radio on. Use the ring to select F1 SETRST (reset settings). Press [**F/L**] to reset.

Yaesu FT-2D

Radio Layout

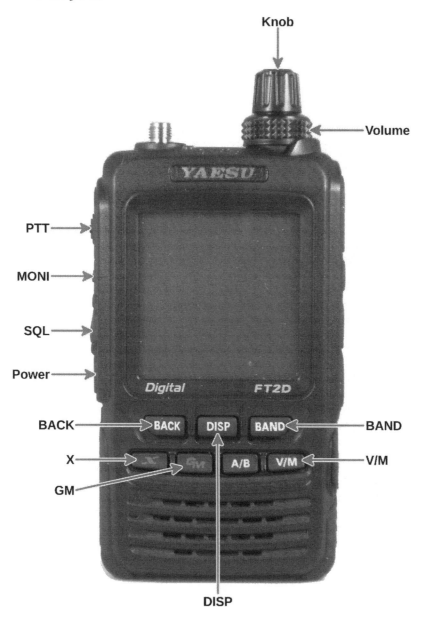

Knob

Volume

PTT

MONI

SQL

Power

BACK

X

GM

BACK

DISP

BAND

BAND

V/M

V/M

DISP

Specs

Receivers Two independent receivers, simultaneous receive
Receives 522–1720 kHz AM/FM, 1.8–774 MHz AM/FM, 803–999 MHz AM/FM
Transmits 144–148 MHz @ 5 W FM/C4FM, 430–450 MHz @ 5 W FM/C4FM
Antenna connector SMA F on radio; needs SMA M antenna
Modes FM, C4FM Fusion
Memory Channels 900
Power 4–14 V DC, EIAJ-02 barrel style, 4mm OD, 1.7mm ID plug, center positive
Model year 2015

Standard Tasks

Program frequency in the field

1. If you aren't already in VFO mode, press [V/M]. The display will show VFO.

2. If needed, change band: press [BAND].

3. Press [DISP] until you see F MW on the screen.

4. Set frequency: tap the frequency on the screen. The radio will then display a soft keyboard which lets you enter frequency. Enter 144390 for 144.390 MHz.

5. Set mode: press [MODE] until you see FM. Other options are DN (Fusion digital), VW (full bandwidth digital voice), and DW (full bandwidth digital data). If you see a mode with a bar over it, that is an automatically selected mode.

6. Set transmit tone type: press F MW and then press [SQ TYPE]. Use the knob to cycle through OFF (no tone), TONE (CTCSS tone on transmit), TONE SQL (CTCSS tone on receive), DCS (digital coded squelch), REV TONE (squelch if tone received), PR FREQ ("no-communication squelch" with a frequency) and PAGER (detect two CTCSS tones). Press [PTT] (does not transmit) to set.

7. Set repeater shift: hold [DISP] for at least one second to enter set mode. Press [CONFIG]. Use the knob to select 14RPT ARS and touch that to turn it on or off. Use the knob to select 15RPT SHIFT and touch that to cycle through -RPT (negative shift), +RPT (positive shift) and SIMPLEX (no shift). Press [PTT] (does not transmit) to set.

8. If you need to change the repeater offset frequency: hold [DISP] for at least one second to enter set mode. Press [CONFIG]. Use the knob to select 16RPT SHIFT FREQ, touch that, and use the knob to adjust. Press [PTT] (does not transmit) to set.

9. Set transmit tone: hold [DISP] for at least one second to enter set mode. Press [SIGNALING]. Use the knob to select 12TONE SQL FREQ and touch that to change. Use the knob to adjust the value. Press [PTT] (does not transmit) to set.

10. Set transmit power: press **F MW** and then **TXPWR**. Use the knob to select from `HIGH` (5 W), `LOW1` (0.1 W), `LOW2` (1 W), and `LOW3` (2.5 W). Use **PTT** (does not transmit) to set.

11. Write to a memory: hold **F MW** for about one second, but release it quickly after that.

12. Use the knob to select the memory to write.

13. Press **M.WRITE** to save to memory.

14. Go to memory mode: press **V/M**.

15. Select the memory you just wrote: use the knob.

Lock/unlock radio

Press **Power** to lock/unlock.

Check repeater input frequency

Press **F MW** and then **REV** to switch to reverse input. (In reverse mode the offset indicator will blink.) Repeat to switch back.

Change power in the field

To set transmit power, press **F MW** and then **TXPWR**. Use the knob to select from `HIGH` (5 W), `LOW1` (0.1 W), `LOW2` (1 W), and `LOW3` (2.5 W). Use **PTT** (does not transmit) to set.

Adjust volume

Rotate the ring to adjust the volume.

Adjust squelch

Press **SQL**. Use the knob to adjust squelch from 0–15. Press **SQL** to set.

Weird Modes

Radio doesn't transmit

This radio has a PTT lock. Unlock the radio to transmit.

Signals are weak

This radio has a built-in signal attenuator. To enable or disable, hold **DISP** to enter set mode, then push **TX/RX**. Press **MODE** and then **ANTENNA ATT**. Use the knob to turn it on/off, and then press **PTT** (does not transmit) to set.

Radio shows GM and some functions don't work

You have activated the Group Monitor mode. Press GM to turn it off.

Radio shows WIRES-X and some functions don't work

You have activated the WIRES-X mode. Hold X to turn it off. Press X to turn it on.

Useful Information

Hold Power for about two seconds to turn off or on.

Factory reset

To reset everything, turn the radio off. Then hold BAND, BACK and DISP while turning the radio on. You will see ALL RESET?. Press OK to reset.

Settings reset

To reset most menu settings (but keep programmed memories), turn the radio off. Then hold BACK and DISP while turning the radio on. You will see SET MODE RESET?. Press OK to reset.

Yaesu FT-3D

Radio Layout

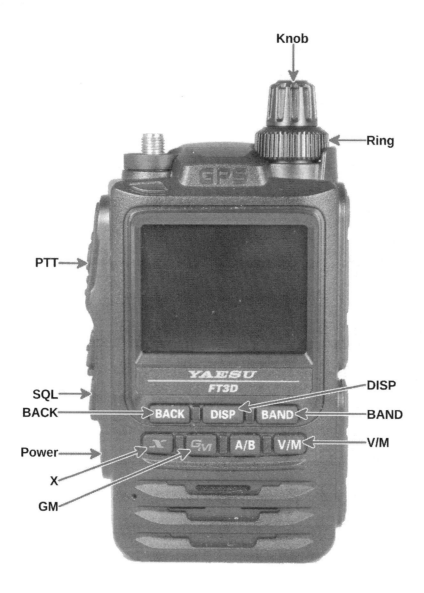

Specs

Receivers Two independent receivers, simultaneous receive
Receives 500 kHz–999.995 MHz (no cell), AM/FM/C4FM
Transmits 144–148 MHz @ 5 W FM/C4FM, 430–450 MHz @ 5 W FM/C4FM
Antenna connector SMA F on radio; needs SMA M antenna
Modes FM, C4FM Fusion
Memory Channels 900
Power 10.5–16 V DC, EIAJ-02 barrel style, 4mm OD, 1.7mm ID plug, center positive
Model year 2019

Standard Tasks

Program frequency in the field

1. If you aren't already in VFO mode, press [V/M]. The screen will show show VFO.

2. Set frequency: tap the frequency on the touchscreen. The radio will then display a soft keyboard which lets you enter frequency. Enter 144390 for 144.390 MHz, then tap ENT. As you're entering the value, it may auto-complete without you having to do anything.

3. Set mode: tap MODE until you see FM (there are multiple Fusion digital options indicated by DN). If you see a mode with a bar over it, that is an automatically selected mode.

4. Set transmit tone type: tap F MW then FWD to get to the second menu, then SQTYPE. Use the knob to cycle through OFF (no tone), TONE TN (CTCSS tone on transmit), TONE SQL TSQ (CTCSS tone on receive), DCS DCS (digital coded squelch), REV TONE RTN (squelch if tone received), PR FREQ PR ("no-communication squelch" with a frequency) and PAGER PAG (detect two CTCSS tones). Press [BACK] to get back out and set the value.

5. Set transmit tone: tap F MW and then CODE. Use the knob to set the value. Press [BACK] to get back out and set the value.

6. Set repeater shift: hold [DISP] for at least one second to enter set mode. Tap CONFIG. Use the knob to select 14 RPT ARS and touch that, then use the knob to turn it on or off. Press [BACK]. Use the knob to select 15 RPT SHIFT and touch that to cycle through -RPT (negative shift), +RPT (positive shift) and SIMPLEX (no shift). Press [BACK] to get back out and set the value.

7. If you need to change the repeater offset frequency: hold [DISP] for at least one second to enter set mode. Tap CONFIG. Use the knob to select menu number 16 RPT SHIFT FREQ, touch that, and use the knob to adjust. Press [BACK] to get back out and set the value.

8. Set transmit power: tap F MW and then TXPWR. Use the knob to select from HIGH, LOW1, LOW2, LOW3. Press [BACK] to get back out and set the value.

9. Write to a memory: hold **F MW** for about one second, but release it quickly after that.

10. Use the knob to select the memory to write. Memories which are displayed in white are empty; memories which have values assigned already are displayed in red.

11. Tap **M.WRITE** to save to memory. You are prompted for a name. Use the keyboard to enter it, or press **V/M** to save without entering a name.

12. Go to memory mode: press **V/M**

13. Select the memory you just wrote: use the knob.

Lock/unlock radio

Press **Power** to lock/unlock.

Check repeater input frequency

Tap **F MW** and then **REV** to switch to reverse input. (In reverse mode the offset indicator will blink.) Repeat to switch back.

Change power in the field

To set transmit power, tap **F MW** and then **TXPWR**. Use the knob to select from HIGH (5 W), LOW1 (0.1 W), LOW2 (1 W), and LOW3 (2.5 W). Press **BACK** to get back out and set the value.

Adjust volume

Rotate the ring to adjust the volume.

Adjust squelch

Press **SQL**. Use the ring to adjust squelch from 0–15. Press **SQL** to set.

Weird Modes

Radio doesn't transmit

This radio has a PTT lock. Unlock the radio to transmit.

Received signals are weak

This radio has a built-in signal attenuator. To enable or disable, hold **DISP** to enter set mode, then tap **TX/RX**. Tap **MODE** and then **ANTENNA ATT**. Use the knob to turn it on/off, and then press **BACK** to set the value and exit. The radio will display ATT when attenuation is enabled.

Radio shows GM and some functions don't work

You have activated the Group Monitor mode. Press GM to turn it off.

Radio shows WIRES-X and some functions don't work

You have activated the WIRES-X mode. Hold X to turn it off. Press X to turn it on.

Knob not working as expected

This radio can exchange the knob with the ring in software. Hold DISP and then tap CONFIG. Use the knob to select item 22 DIAL KNOB CHANGE and touch that. Tap CHANGE to swap the values. Press BACK to get back out and set the value.

Useful Information

Hold Power for about two seconds to turn the radio off or on.

Factory reset

To reset everything, turn the radio off. Then hold BAND, BACK and DISP while turning the radio on. You will see ALL RESET?. Tap OK to reset.

Settings reset

To reset most menu settings (but keep programmed memories), turn the radio off. Then hold BACK and DISP while turning the radio on. You will see SET MODE RESET?. Tap OK to reset.

Yaesu FT-4XR

Radio Layout

Power/
Volume

PTT

MONI

F

▲

▼

P2

*

#

Specs

Receivers Single receiver with priority channel option
Receives 65–108 MHz WFM, 136–174 MHz FM, 400–480 MHz FM
Transmits 144–148 MHz @ 5 W FM, 400–480 MHz @ 5 W FM
Antenna connector SMA F on radio; needs SMA M antenna
Modes FM
Memory Channels 200
Power No DC input on radio, nominally 7.4 V DC
Model year 2018

Standard Tasks

Program frequency in the field

1. If you aren't already in VFO mode, press `*` (V/M) repeatedly. The radio will cycle through VFO A, VFO B, and memory mode. You should see `tun` when you're in the right mode.

2. If needed, change band: press `#` (BAND) repeatedly to cycle through bands: 2 m, 70 cm and FM broadcast (VFO B, receive only).

3. Set frequency: use the keypad (144390 for 144.390 MHz).

4. If needed, set step size: hold `F` for one second to enter set mode. Press `▲`/`▼` to navigate to menu `37 STEP`. Press `F` to change. Use `▲`/`▼` to set the desired step size. Press `PTT` (does not transmit) to save.

5. Set repeater shift: hold `F` for one second to enter set mode. Press `▲`/`▼` to navigate to menu `31 RPT.SFT`. Press `F` to change. Use `▲`/`▼` to select from `SIMPLX` (simplex), `+RPT` (positive offset), or `-RPT` (negative offset). Press `PTT` (does not transmit) to save.

6. If you need to change the repeater offset frequency: hold `F` for one second to enter set mode. Then use `▲`/`▼` to navigate to menu `30 RPT.FRQ`. Press `F` to change. Use `▲`/`▼` to change the value. (The keypad does not appear to work for changing the offset frequency.) Press `PTT` (does not transmit) to set.

7. Set transmit tone type: hold `F` for one second to enter set mode. Then use `▲`/`▼` to navigate to menu `36 SQL.TYPE`. Press `F` to change. Use `▲`/`▼` to cycle through `OFF` (no tones), `R-TONE` (require receive tone only, no transmit tone), `T-TONE` (transmit tone only, no receive), `TSQL` (transmit tone and require receive tone), `REV TN` (receive only when tone is *not* present), `DCS` and `PAGER` (two-tone CTCSS calling). Press `PTT` (does not transmit) to set.

8. Set transmit tone: hold `F` for one second to enter set mode. Then use `▲`/`▼` to navigate to menu `38 TN FRQ`. Press `F` to change. Use `▲`/`▼` to change the value. (The keypad does not appear to work for changing the tone frequency.) Then press `PTT` (does not transmit) to set. You must set tone type before setting tone value, otherwise this menu will not be enabled.

9. Set transmit power: hold ⬚F⬚ for one second to enter set mode. Then use ⬚▲⬚/⬚▼⬚ to navigate to menu `40 TX PWR`. Press ⬚F⬚ to change. Use ⬚▲⬚/⬚▼⬚ to select power level. Options are `HIGH` (5 W)/`MID` (2.5 W)/`LOW` (0.5 W). Press ⬚PTT⬚ (does not transmit) to set.

10. Write to a memory: hold ⬚*⬚(V/M) for at least one second to go to memory mode.

11. Press ⬚▲⬚ ⬚▼⬚ to select the desired memory.

12. You are presented with a second entry field. You can use this to set the name with the keypad and ⬚F⬚ keys.

13. Hold ⬚*⬚(V/M) to write the memory.

14. Go to memory mode: press ⬚*⬚(V/M).

15. Select the memory you just wrote: use ⬚▲⬚⬚▼⬚.

Lock/unlock radio

Hold ⬚6⬚ to lock/unlock. Menu `18 LOCK` lets you select what gets locked.

Check repeater input frequency

Press ⬚F⬚ then ⬚P2⬚ to switch between reverse and regular modes. (Repeater shift icon will blink in reverse mode.) Press ⬚F⬚ then ⬚P2⬚ to go back to normal.

Change power in the field

Hold ⬚F⬚ for one second to enter set mode. Then use ⬚▲⬚⬚▼⬚ to navigate to menu `40 TX PWR`. Press ⬚F⬚ to change. Use ⬚▲⬚⬚▼⬚ to select power level. Options are `HIGH` (5 W)/`MID` (2.5 W)/`LOW` (0.5 W). Press ⬚PTT⬚ (does not transmit) to set.

Adjust volume

Rotate the volume power/volume knob.

Adjust squelch

Press ⬚F⬚ then ⬚MONI⬚. Use ⬚▲⬚⬚▼⬚ to adjust the squelch from `LVL 0` through `LVL 15`. Press ⬚F⬚ to save.

Weird Modes

Can't set repeater shift

This radio has an automatic repeater shift. To disable it, hold ⬚F⬚ to enter set mode. Press ⬚▲⬚⬚▼⬚ to navigate to menu `29 RPT.ARS`. Press ⬚F⬚ to change. Use ⬚▲⬚⬚▼⬚ to set `ARS.OFF` to disable ARS, or `ARS. ON` to enable. Press ⬚PTT⬚ (does not transmit) to save.

Can't receive signals even with squelch all the way down

This radio has an RF Squelch option, which, if enabled, will require a signal of a certain S-level to be present before opening the squelch. To change it, hold [F] to enter set mode. Press [▲][▼] to navigate to menu 28 RF SQL. Press [F] to change. Use [▲][▼] to select OFF (or a value of your choosing, S-1 through S6, S-8 or S-FULL). Press [PTT] (does not transmit) to save.

Radio shows P before frequency, or shows PAGING

The two CTCSS tone pager squelch type is enabled. Change squelch type to turn it off.

Radio doesn't transmit

This radio has a PTT lock. Unlock the radio to transmit.

Can't enter VFO mode

This radio has a memory-only mode. To disable it, turn the radio off. Hold [MONI] and [PTT] (does not transmit) while turning it on. Use [▲][▼] to navigate to F5 M-ONLY. Press [F] to leave (or enter) this mode.

Useful Information

You can change the lock setting by holding [F] to enter set mode, then navigating with [▲][▼] to 18 LOCK. Press [F] to change. Use [▲][▼] to choose from LK KEY (keyboard lock), LK PTT (PTT lock), and LK P+K (lock both keyboard and PTT). Press [PTT] (does not transmit) to save.

The FT-4V is the same radio as the FT-4XR, but is 2 m only rather than dual-band. It is programmed similarly. You can tell the difference by the keypad, which will have [#](VFO) and [*](MR) instead of [*](V/M) and [#](BAND). On the FT-4V, press [#] (VFO) to enter VFO mode, press [*](MR) to enter memory mode, and hold [*](MR) to write a memory.

To enable expanded transmit range (137–174 MHz and 420–470 MHz) turn the radio off. Hold [MONI] and [PTT] (does not transmit). You will see F1 SET.RST. Using the keypad, enter 32406665 (some versions require 22406665 or 62406665 instead). The radio will reset reset. At this point you can transmit out of band. Repeat the procedure to go back to the default.

Factory reset

Turn the radio off. Hold [MONI] and [PTT] (does not transmit) while turning it on. Use [▲][▼] to select F4 ALL.RST. Press [F] to reset.

Memory reset

Turn the radio off. Hold [MONI] and [PTT] (does not transmit) while turning it on. Use [▲][▼] to select F2 MEM.RST. Press [F] to reset.

Memory bank reset

Turn the radio off. Hold [MONI] and [PTT] (does not transmit) while turning it on. Use
[▲][▼] to select F3 MB.RST. Press [F] to reset.

Settings reset

Turn the radio off. Hold [MONI] and [PTT] (does not transmit) while turning it on. Use
[▲][▼] to select F1 SET.RST. Press [F] to reset.

Yaesu FT-50

Radio Layout

Specs

Receivers Single receiver with priority channel option (memory 1 in memory mode)
Receives 76–200 MHz AM/FM, 300–400 MHz AM/FM, 400–540 MHz AM/FM, 590–999 MHz AM/FM
Transmits 144–148 MHz @ 5 W FM, 430–450 MHZ @ 5 W FM
Antenna connector SMA F on radio; needs SMA M antenna
Modes FM
Memory Channels 99
Power 5–13 V DC, EIAJ-02 barrel style, 4mm OD, 1.7mm ID plug, center positive
Model year 1996

Standard Tasks

Program frequency in the field

1. If you aren't already in VFO mode, press **VFO**. The radio has two VFO modes (A and B); you can be in either. The screen will show `CH-` if you aren't in VFO mode.

2. If needed, set step size: press **F** then **7** to adjust step size, then rotate knob to select the desired `STEP` value. Press **7** to exit.

3. Set frequency: Use the keypad (144390 for 144.390 MHz). If step size is greater than 5, you might not have to/be able to enter the last digit.

4. The radio has automatic repeater shift (ARS). To override that, hold **KNOB** for half a second, use the knob to select `RPTR -06-` and press **KNOB**. Rotate to select from `-RPT` (negative shift), `SIMP` (simplex) and `+RPT` (positive shift). Press **PTT** (does not transmit) to leave menu and save shift.

5. Set transmit tone type: press **TN** to cycle through `T` (tone), `T SQ` (tone squelch—only if FTT-12 is installed) and `DCS` (DCS). Press **F** then **TN** to enter tone adjust mode. Use the knob to select tone, then press **TN**.

6. Set transmit power: hold **KNOB** for half a second, then use the knob to select `TXPO-02-`. Press **KNOB** and use the knob to select power level: `HI` 5 W / `L1` 0.1 W / `L2` 1 W / `L3` 2.8 W. Press **PTT** (does not transmit) to exit.

7. Write to a memory: hold **F** for at least half a second to start write.

8. Use the knob to select desired memory. Memories which are already programmed appear as `CH-2`; empty memories appear as `CHu2`.

9. Press and release **F** to write.

10. Go to memory mode: press **MR**.

11. Select the memory you just wrote: use the knob.

Lock/unlock radio

Press [F] then [LW] to lock/unlock. The type of lock/unlock can be changed with menu `LOCK -17-`. Lock options are keypad `KL`, knob `DL`, PTT `PL`, and combinations of those.

Check repeater input frequency

Press and release [RV] to switch between reverse and regular modes.

Change power in the field

To set transmit power, hold [KNOB] for half a second, then use the knob to select `TXPO-02-`. Press [KNOB] and use the knob to select power level: `HI` 5 W / `L1` 0.1 W / `L2` 1 W / `L3` 2.8 W. Press [PTT] (does not transmit) to exit.

Adjust volume

Rotate the outer volume ring to adjust volume.

Adjust squelch

Hold [KNOB] for half a second. Scroll to `SQL -01-`. Press [KNOB] and select squelch level (0–15) with the knob. Press [PTT] (does not transmit) to exit.

Weird Modes

If someone says you have WIRES mode turned on

This radio doesn't have WIRES, but does have a DTMF paging mode. To exit it, press [F] and then repeatedly press [1] until you don't see `CODE` or `PAGE` on the bottom of the screen. Then press [F].

Radio shows TRX, TX or RX

This means you have ARTS mode enabled. To disable, press [F] then hold [KNOB] for half a second. Press [F] then press [TN]. Finally, press [F] repeatedly until you no longer see `TRX`, `TX` or `RX`. Then press [TN].

Can't enter VFO mode

This radio has a memory-only mode. To exit it (or enter it again), turn the radio off. Then hold [PTT] and [LAMP] and turn the radio on again.

Radio shows all LCD segments

This radio has an LCD test mode, entered by turning on the radio while holding [LAMP]. Turn the radio off and then back on to clear.

Automatic repeater shift is wrong

There is a modification to enable cell band receive for this radio. If that modification has been applied, ARS will be incorrect for North America (it will be European shifts) and offsets will be wrong. You will have to set the repeater shifts and offsets manually.

Useful Information

Hold **Power** for half a second to turn the radio on or off.

Some versions of this radio have the FTT-12 option installed. This enables brief recordings as well as tone squelch mode. Radios with this option installed have an "R" marking to the left of **TN** and a "P" marking to the left of **RV** on the keypad.

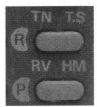

FTT-12 installed   FTT-12 not installed

To turn the radio on or off, hold **Power** for about one second.

The radio has an extended receive mode. To enable or disable it, hold **KNOB** and **LAMP** while powering on.

The radio has a game mode. To enter it, hold **KNOB** and **MR** while powering on. Select speed with the knob, then press **PTT** (does not transmit) to start. A number will display and scroll to the right. Enter the number which will cause the total to be 10, then press **F** (use **1** **0** for 0). Turn the radio off to exit.

This radio has a service mode which can be entered by holding **KNOB**, **PTT** and **LAMP** while turning on the radio. Be cautious since it is possible to convert a working radio into a non-working radio in this mode. Turn the radio off to exit the mode.

Factory reset

To reset everything, hold **MONI** and **KNOB** then turn the radio on. Display will show `ALRST PrS F`. Press **F** to reset.

Radio Layout

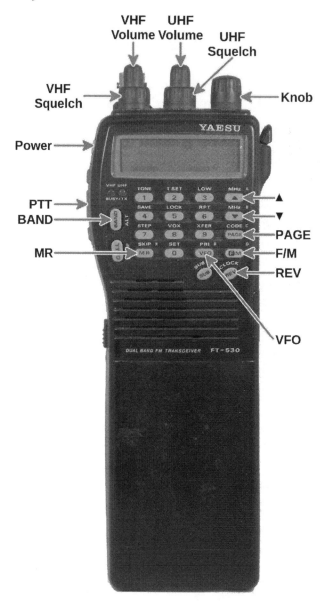

VHF Volume   UHF Volume

UHF Squelch

VHF Squelch

Knob

Power

PTT

BAND

MR

▲

▼

PAGE

F/M

REV

VFO

Specs

Receivers Two independent receivers, simultaneous receive
Receives 130–174 MHz AM/FM, 430–450 MHz FM
Transmits 144–148 MHz @ 5 W FM, 430–450 MHZ @ 5 W FM
Antenna connector BNC F on radio; needs BNC M antenna
Modes FM
Memory Channels 76 total—38 on left side, 38 on right side
Power 5–13 V DC, EIAJ-02 barrel style, 4mm OD, 1.7mm ID plug, center positive
Model year 1992

Standard Tasks

Program frequency in the field

1. If needed, change band: press **BAND** to select either the left band (2 m) or the right band (70 cm).

2. If you aren't already in VFO mode, press **VFO**. The screen will show no channel number in VFO mode. There are two VFO modes (A and B); you can be in either. o

3. Set frequency: use the keypad (144390 for 144.390 MHz).

4. If you need to change the repeater offset frequency: press **F/M** **0** **6**. Use the knob to set. Press **6** to save. Note this applies to everything on the band.

5. Set repeater shift: press **F/M** then repeatedly press **6** to cycle through -, + and blank (no offset). Press **F/M** to set.

6. Set transmit tone: press **F/M** then **2**. Use the knob to select tone. Press **2** to set.

7. Set transmit tone type: first press **F/M** then repeatedly press **1** to cycle through T (tone), T SQ (tone squelch) and blank (no tone). Press **F/M** to set.

8. Set transmit power: press **F/M** then **3**. You will see either Hi or something that starts with L. Hi corresponds to high power (5 W on 12 V DC, 2 W on 7.2 V DC). If you are on L, you can use the knob to select from L1 (0.5 W), L2 (1.5 W) or L3 (2 W). Wait two seconds after setting level to go back to normal mode. The display will show LOW when you are using a low power setting, and nothing when you are using the high power setting.

9. Write to a memory: hold **F/M** for at least half a second.

10. Use the knob to select the desired memory. Memories with data display as solid; memories without data display blinking.

11. Press and release **F/M** to write.

12. Go to memory mode: press **MR**.

13. Select the memory you just wrote: use the knob.

Lock/unlock radio

Press [F/M] then [5] repeatedly to lock PTT (PL), keyboard (KL) and knob (DL if enabled). Press [F/M] or wait two seconds to set.

Check repeater input frequency

Press and release [REV] to switch between reverse and regular modes.

Change power in the field

To set transmit power, press [F/M] then [3]. You will see either Hi or something that starts with L. Hi corresponds to high power (5 W on 12 V DC, 2 W on 7.2 V DC). If you are on L, you can use the knob to select from L1 (0.5 W), L2 (1.5 W) or L3 (2 W). Wait two seconds after setting level to go back to normal mode. The display will show LOW when you are using a low power setting, and nothing when you are using the high power setting.

Adjust volume

Rotate the center volume knob for the correct side of the radio to adjust volume.

Adjust squelch

Rotate the outer squelch ring for the correct side of the radio to adjust squelch.

Weird Modes

Radio shows blinking LOW

The radio will display a blinking LOW if it is overheating. Stop transmitting/move to a lower power level.

Received audio bad

The radio can be switched into AM mode for 2 m. To enable/disable AM, press [F/M] [0] then repeatedly press [F/M] [VFO] until you see A3 on (AM on) or A3 OFF (AM off). Press [0] to set.

Radio plays annoying tones when you press any button

This can be disabled/enabled with [F/M] [2] [F/M] [2].

Radio plays tones when keyed up (PTT)

The radio has a number of page/code modes. Repeatedly press [PAGE] until you don't see a bell symbol, CODE or PAGE in the display.

Useful Information

Press **Power** to turn the radio on. Hold it for half a second to turn the radio off.

This radio has crossband repeat capability. (This is full crossband; whatever is transmitted on one frequency will be sent out on the other. Turn the volume knobs down to avoid feedback.) To enable, press **6** while turning on the radio. Whatever frequencies/tones that were in memory for VHF and UHF will be used to crossband repeat.

The radio will transmit VHF only on the left display, and transmit UHF only on the right display. You can receive either frequency on either display.

This radio has a clock. The clock display can be turned on or off with **F/M** **REV** **5** **REV**.

The radio has automatic repeater shift (ARS). It is disabled by default; to toggle it press **F/M** **0** **6** **F/M**. You will see **A** when ARS is turned on. Press **6** to exit.

The radio has an extended receive mode that can be enabled after a hardware modification. If the modification has been done, you can enter extended receive by holding down **▲** and **▼** while powering on. After receive has been extended, performing this procedure will reset everything. After entering extended receive, program 110.000 in the left side L memory, 180.000 in the left side U memory, 300.000 in the right side L memory and 500.00 in the right side U memory.

Factory reset

To reset everything, hold **MR** and **VFO** then turn the radio on. This will reset everything (including channel memories) without any prompts.

Yaesu FT-60

Radio Layout

Specs

Receivers Single receiver, dual watch (first to break squelch wins)
Receives 108–520 MHz AM/FM, 700–999 MHz AM/FM
Transmits 144–148 MHz @ 5 W FM, 430–450 MHZ @ 5 W FM
Antenna connector SMA F on radio; needs SMA M antenna
Modes FM
Memory Channels 1000
Power 6–16 V DC, EIAJ-02 barrel style, 4mm OD, 1.7mm ID plug, center positive
Model year 2004

Standard Tasks

Program frequency in the field

1. If you aren't already in VFO mode, press [V/M]. The screen will show channel numbers in the upper left if you aren't in VFO mode.

2. Set frequency: use the keypad (144390 for 144.390 MHz).

3. Enter menu mode: press [F/W][0](SET).

4. Set transmit tone type: use the knob to select `48:SQL.TYP`. Press [F/W] to adjust, and use the knob to select from `TONE` (transmit tone, no tone squelch), `TSQL` (transmit tone and receive tone), `REV TN` (squelch when tone received), `DCS` (DCS) or `OFF`. Press [F/W] when done.

5. Set transmit tone: use the knob to select `50:TN FRQ`. Press [F/W] to adjust tone frequency. Use the knob to adjust, and press [F/W] when done. Use menu `13:DCS.COD` instead for DCS.

6. Set repeater shift: use the knob to select `38:RPT.MOD`. Press [F/W] to adjust, and use the knob to select from `RPT. +`, `RPT.OFF` (simplex) or `RPT. -`. Press [F/W] when done.

7. Exit menu mode: hold [F/W] for at least half a second.

8. Set transmit power: press [F/W][3](TX PO). Use the knob to select power level: `HIGH` 5 W, `MID` 2 W, or `LOW` 0.5 W. Press and release [F/W] to set.

9. Write to a memory: hold [F/W] for at least half a second.

10. Use the knob to select the desired memory.

11. Press and release [F/W] to write.

12. Go to memory mode: press [V/M]. The display will show `MEMORY` briefly and then show a channel number in the upper left.

13. Select the memory you just wrote: use the knob.

Lock/unlock radio

Press **F/W** **6** (LOCK) to lock/unlock.

Check repeater input frequency

Press and release **HM/RV** to switch between reverse and regular modes.

Change power in the field

To set transmit power, press **F/W** **3** (TX PO). Use the knob to select power level: HIGH 5 W, MID 2 W, or LOW 0.5 W. Press and release **F/W** to set.

Adjust volume

Rotate the center power/volume knob to adjust volume.

Adjust squelch

Rotate the outer ring of the selection knob to adjust RF squelch.

Weird Modes

If someone says you have WIRES mode turned on

You are transmitting a DTMF tone that interrupts your first few syllables. You will see ⊗ in the upper right of screen. Press and release the **0** (SET) key once. The ⊗ should go away.

Radio shows OUTRNG

This means you have ARTS mode enabled. Press and hold **4** (CODE) for two seconds to disable.

Can't enter VFO mode

This radio has a "memory-only" mode which prevents you from entering frequencies directly. To disable it (or re-enable it), hold **MONI** and turn the radio on. Use the knob to select F5:M-ONLY. Press **F/W**. The radio will reboot.

Can't enter VHF or UHF frequencies

This radio can be locked to UHF only or VHF only. To disable (or re-enable) this mode, hold **MONI** and turn the radio on. Use the knob to select F6:V-ONLY or F7:U-ONLY. The radio will reboot.

DTMF doesn't send tone on transmit

By default, the radio starts up in DTMF Memory mode. Pressing a number key while you hold [PTT] will send the number stored in that memory. Usually these memories are empty so nothing will be sent, although [*] and [#] always transmit.

   To have the radio play the appropriate touchtone instead, press [F/W] [9] (DTMF). The radio will show CODE. At that point, pushing a number while [PTT] is held will send the touch tone. Press [F/W] [9] again to switch back to memory mode (MEM), if desired.

Useful Information

There are two programmable function keys, [7] (P1) and [8] (P2). By default, [7] (P1) is assigned to 29:PAGER and [8] (P2) is assigned to 46:SKIP. These can be changed: enter memory mode with [F/W] [0] (SET) and then select a menu using the knob. Next, hold [7] (P1) or [8] (P2) for one second to assign the menu item to that button. Hold [F/W] to exit the menu.

Factory reset

To reset everything, hold [MONI] and turn the radio on. Use the knob to select F4:ALLRST (reset everything). Press [F/W] to reset.

Memory reset

To reset all memories, hold [MONI] and turn the radio on. Use the knob to select F2:MEMRST (reset memories). Press [F/W] to reset.

Memory bank reset

To reset memory bank assignments but not memories, hold [MONI] and turn the radio on. Use the knob to select F3:MB RST (reset memory banks). Press [F/W] to reset.

Settings reset

To reset most settings but leave memories intact, hold [MONI] and turn the radio on. Use the knob to select F1:SETRST (reset settings). Press [F/W] to reset.

Radio Layout

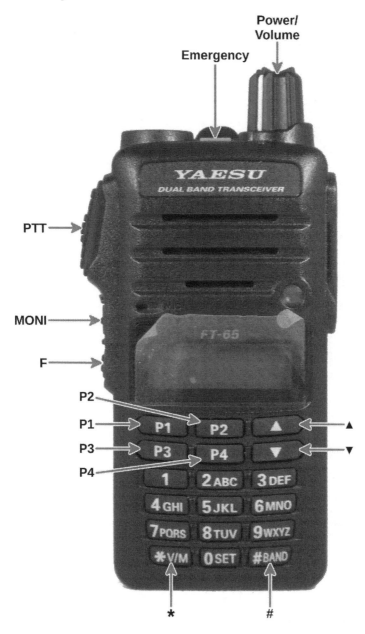

Power/
Volume

Emergency

PTT

MONI

F

P2

P1

P3

P4

▲

▼

*

#

Specs

Receivers Single receiver with priority channel option
Receives 65–108 MHz WFM, 136–174 MHz FM, 400–480 MHz FM
Transmits 144–148 MHz @ 5 W FM, 430–450 MHz @ 5 W FM
Antenna connector SMA **M** on radio; needs SMA **F** antenna
Modes FM
Memory Channels 200
Power No power input on radio, nominally 7.4 V DC
Model year 2017

Standard Tasks

Program frequency in the field

1. If you aren't already in VFO mode, press ⟦ * ⟧. The screen shows `M` if you're in memory mode.

2. Set frequency: use the keypad (144390 for 144.390 MHz).

3. Enter menu mode: hold ⟦ F ⟧.

4. Set transmit tone: use ⟦▲⟧⟦▼⟧ to scroll to item 8, `CTCSS`. Press ⟦ F ⟧ to select `TX` if desired, then ⟦▲⟧⟦▼⟧ to adjust. Press ⟦ F ⟧ to set. Similarly, use ⟦▲⟧⟦▼⟧ to select `RX` if desired, then ⟦▲⟧⟦▼⟧ to adjust. Press ⟦ F ⟧ to set.

5. Set repeater shift: use ⟦▲⟧⟦▼⟧ to scroll to item 24, `Repeater`. Press ⟦ F ⟧ to select `Mode`. Use ⟦▲⟧⟦▼⟧ to switch between `Simplex`, `+REP` (positive offset) and `-REP` (negative offset). Press ⟦ F ⟧ to set.

6. If you need to change the repeater offset frequency: use ⟦▼⟧ to select item 25, `Shift`. Press ⟦ F ⟧ to change, and use ⟦▲⟧⟦▼⟧ to adjust. Press ⟦ F ⟧ to set when done.

7. Set transmit tone type: use ⟦▲⟧⟦▼⟧ to scroll to item 29, `SQL Type`. Press ⟦ F ⟧ to adjust. Use ⟦▲⟧⟦▼⟧ to select from `R-TONE` (receive squelch), `T-TONE` (transmit tone), `TSQL` (transmit tone and receive tone squelch), `REV TN` (squelch if tone received), `DCS` (digital coded squelch), `PAGER` (unsquelch if two CTCSS tones received) , and `OFF` (no tone). Press ⟦ F ⟧ to set.

8. Set transmit power: use ⟦▲⟧⟦▼⟧ to go to item 32, `TX PWR`. Press ⟦ F ⟧ to adjust. Use ⟦▲⟧⟦▼⟧ to choose from `HI(5 W)`, `MID(2.5 W)` or `LOW(0.5 W)`. Press ⟦ F ⟧ to set.

9. Exit menu mode: hold ⟦ F ⟧.

10. Write to a memory: hold ⟦ * ⟧.

11. Use ⟦▲⟧⟦▼⟧ to select the memory.

12. hold ⟦ * ⟧ to complete writing to memory.

13. Go to memory mode: press [*].

14. Select the memory you just wrote: use [▲][▼].

Lock/unlock radio

Hold [6] to lock/unlock.

Check repeater input frequency

Press [F][P4] to enter or exit reverse mode. Shift will blink when you're in reverse mode.

Change power in the field

To set transmit power, hold [F] to enter the menu. Use [▲][▼] to go to item 32, TX PWR. Press [F] to adjust. Use [▲][▼] to choose from HI(5 W), MID(2.5 W) or LOW(0.5 W). Press [F] to set. Hold [F] to exit the menu.

Adjust volume

Rotate the knob to adjust volume.

Adjust squelch

Hold [F] to enter the menu. Scroll to item 26, RF SQL. Press [F] to adjust. Use [▲][▼] to go from OFF, S-1 through S-8, or S-FULL. Press [F] to set. Hold [F] to exit the menu.

Weird Modes

Radio is beeping continually

You are in emergency mode. Turn the radio off and then on. If you hold down [Emergency] for three seconds, you will enter emergency mode. This causes a continual beep and wil transmit on the home frequency.

Can't enter VFO mode

This radio has a memory-only mode. To exit, turn the radio off. Then hold [MONI] and [PTT] (does not transmit) while turning the radio on. Use [▼] to select F5:MEM-ONLY. Press [F] to toggle.

Can't enter VHF or UHF frequencies

This radio has an option to use VHF only or UHF only. To adjust it, turn the radio off. Then hold [MONI] and [PTT] (does not transmit) while turning the radio on. Use [▼] to select F6:VHF-ONLY or F7:UHF-ONLY. Press [F] to toggle.

Radio shows OUT.RNG or IN.RNG

This means you have ARTS mode enabled. Hold [2] to exit.

Radio shows - - - - on power on

The password has been enabled. Enter the password (four digits) on the keypad. If you have forgotten the password, you can disable with a full reset (see below).

Useful Information

This radio has four programmable buttons which remember the radio state. Rather than programming a memory, once you've set up the frequency, hold [P1], [P2], [P3] or [P4] to program the button. Then later you can press [P1], [P2], [P3] or [P4] to recall the state (including frequency/tone info).

To enable expanded transmit range (137–174 MHz) turn the radio off. Then hold down [MONI] and [PTT] (does not transmit) while turning the radio on. You will see F1 SET RESET. Using the keypad, enter 32406665 (some versions require 22406665 or 62406665 instead). The radio will reset reset. At this point you can transmit out of band. Repeat the procedure to go back to the default.

Factory reset

To reset, hold [MONI] and [PTT] (does not transmit) while turning the radio on. Use [▲]/[▼] to select F4 ALL RESET (full reset—reset everything). Press [F] to reset.

Memory reset

To reset, hold [MONI] and [PTT] (does not transmit) while turning the radio on. Use [▲]/[▼] to select F2 MEM RESET (reset memories). Press [F] to reset.

Memory bank reset

To reset, hold [MONI] and [PTT] (does not transmit) while turning the radio on. Use [▲]/[▼] to select F3 BANK RESET (reset memory banks). Press [F] to reset.

Settings reset

To reset, hold [MONI] and [PTT] (does not transmit) while turning the radio on. Use [▲]/[▼] to select F1 SET RESET (reset settings). Press [F] to reset.

Radio Layout

Knob

PTT

MONI

GM

VOL

F

Power

MODE

HM/RV

AMS

BAND

V/M

Specs

Receivers Single receiver with priority channel option
Receives 108–136.995 MHz AM, 137–579.995 MHz FM, 144–147.995 MHz C4FM, 430–449.995 C4FM
Transmits 144–147.995 MHz @ 5 W FM/C4FM, 430–449.995 MHz @ 5 W FM/ C4FM
Antenna connector SMA F on radio; needs SMA M antenna
Modes FM, C4FM Fusion
Memory Channels 999
Power 6.0–16.0 V DC, nominally 7.4 V DC, EIAJ-02 barrel style, 4mm OD, 1.7mm ID plug, center positive
Model year 2017

Standard Tasks

Program frequency in the field

1. If you aren't already in VFO mode, press [V/M]. The screen will show channel number if you aren't in VFO mode.

2. If needed, change band: press [BAND] repeatedly to cycle through bands and select the appropriate one.

3. Set mode: press [MODE] (FM for analog FM mode, DN for Fusion digital mode).

4. If needed, set step size: press [F] [4] (STEP). Use the knob to adjust the step size. Press [PTT] (does not transmit) to set.

5. Set frequency: use the keypad (144390 for 144.390 MHz).

6. Set repeater shift: press [F] [0] (RPT) to cycle through RPT- (negative offset), RPT+ (positive offset) and SIMP (simplex).

7. If you need to change the repeater offset frequency: hold [F] for one second. Then use the knob to select menu 46 RPT.FRQ. Press [F] to change, use the knob to adjust, then press [PTT] (does not transmit) to set.

8. Set transmit tone type: press [F] [5] (SQ TYP) repeatedly to cycle through OFF (no tone), TONE (send tone on transmit), TSQL (tone on transmit and receive), DCS (digital coded squelch), RV TN (reverse tone—squelch if tone received), PR FRQ ("no-communication squelch" based on a tone), PAGER (play bell if two CTCSS tones received). Press [PTT] (does not transmit) to set.

9. Set transmit tone: press [F] [6] (CODE). Use the knob to adjust, and press [PTT] (does not transmit) to set.

10. Set transmit power: press [F] [1] (TX PO) repeatedly to select power level. Options are HIGH (5 W), MID (2 W), or LOW (0.5 W). Press [PTT] (does not transmit) to set.

11. Write to a memory: hold [V/M] for at least one second to go to memory mode.

12. Use the knob to select the memory to write.

13. Press [V/M] to save the frequency.

14. You are presented with a second entry field. You can use this to set the name with the knob and [BAND] keys.

15. Hold [V/M] to write.

16. Go to memory mode: press [V/M].

17. Select the memory you just wrote: use the knob to select the desired memory.

Lock/unlock radio

Presss [Power] to lock/unlock. Menu 30 LOCK lets you set what gets locked.

Check repeater input frequency

Press and release [HM/RV] to switch between reverse and regular modes.

Change power in the field

To set transmit power, press [F] [1] (TX PO) repeatedly to select power level. Options are HIGH (5 W), MID (2 W), or LOW (0.5 W). Press [PTT] (does not transmit) to set.

Adjust volume

Hold [VOL] down and rotate the knob. Release [VOL].

Adjust squelch

Press [F] [MONI]. Use the knob to adjust squelch. Press [PTT] (does not transmit) to set.

Weird Modes

Radio shows GROUP and DG-ID

The Group Monitor function is on. Press [GM] to turn it off again.

Radio doesn't transmit

This radio has a PTT lock. Unlock the radio to transmit.

You can change the lock setting by holding [F] to enter set mode, then selecting memory 30 LOCK. Press [F] to change, and use the knob to select the lock type you want.

Radio shows M-ONLY and can't enter VFO mode

This radio has a memory channel only mode. To disable it (or enable it), turn the radio off. Then hold [V/M] while turning the radio back on.

Useful Information

Hold [Power] for about two seconds to turn off or on.

Additional ways to send tones can be unlocked with menu 54 SQL.EXP.

Factory reset

Start by turning off the radio. To reset everything, hold [MODE], [HM/RV] and [AMS] while turning the receiver on. Then press [F] to reset.

Settings reset

Start by turning off the radio. To reset to default settings but save memories, hold [MODE] and [V/M] while turning on. Then press [F] to reset.

Yaesu VX-150

Radio Layout

Power/Volume

Knob

Ring

PTT

LAMP

MR

VFO

REV

F

Specs

Receivers Single receiver
Receives 137–174 MHz FM
Transmits 144–148 MHz @ 5 W FM
Antenna connector SMA F on radio; needs SMA M antenna
Modes FM
Memory Channels 199
Power 13.8 V DC, EIAJ-02 barrel style, 4mm OD, 1.7mm ID plug, center positive
Model year 2000

Standard Tasks

Program frequency in the field

1. If you aren't already in VFO mode, press `VFO`. The screen will show a channel number if you're not in VFO mode.

2. Set frequency: use the keypad (144390 for 144.390 MHz).

3. Enter menu mode: press `F` `0`.

4. Set transmit tone: use the knob to select menu `26:TN SET` or menu `27:DCS SET`. Press `F` to adjust value. Use the knob to select the desired value.

5. Set transmit tone type: use the knob to select menu `25:SQL TYP.` and press `F` to adjust. There are four settings: `TN ENC` to send tone on transmit, `TN SQL` to send tone on transmit and require tone on receive, `DCS` and `OFF`. Use the knob to select the desired type, and press `F` when done.

6. Set repeater shift: use the knob to select menu `3:RPT`. Press `F` to adjust. Use knob to select from `SIMP` (simplex), `-RPT` (negative) or `+RPT` (positive). Press `F` when done.

7. Exit menu mode: press `PTT` (does not transmit).

8. Set transmit power: press `F` `3`. Use the knob to select the desired output power (`HIGH` 5 W, `MID` 2 W, `LOW` 0.5 W). Press and release `F` to set.

9. Write to a memory: hold `F` for at least one second.

10. Use the knob to select the destination memory within five seconds.

11. Press and release `F` to write.

12. Go to memory mode: press `MR`.

13. Select the memory you just wrote: use the knob.

Lock/unlock radio

Press `F` `6` to lock/unlock.

Check repeater input frequency

Press and release **REV** to switch between reverse and regular modes.

Change power in the field

To set transmit power, press **F** **3**. Use the knob to select the desired output power (HIGH 5 W, MID 2 W, LOW 0.5 W). Press and release **F** to set.

Adjust volume

Rotate the center power/volume knob to adjust volume.

Adjust squelch

Rotate the outer ring of the selection knob to adjust squelch.

Weird Modes

Radio shows OUT RNG

This means you have ARTS mode enabled. Press **F** to exit ARTS mode.

Useful Information

F **0** menu 37:BATT **F** will show battery info.

Factory reset

To reset the radio to factory settings, hold **LAMP** and **PTT** (does not transmit) while turning the radio on. Scroll to ALL.RST (reset settings and memories). Press **F** to reset.

Settings reset

To reset the radio settings only but keep the memories, hold **LAMP** and **PTT** (does not transmit) while turning the radio on. Scroll to SET.RST (reset settings). Press **F** to reset.

Yaesu VX-170

Radio Layout

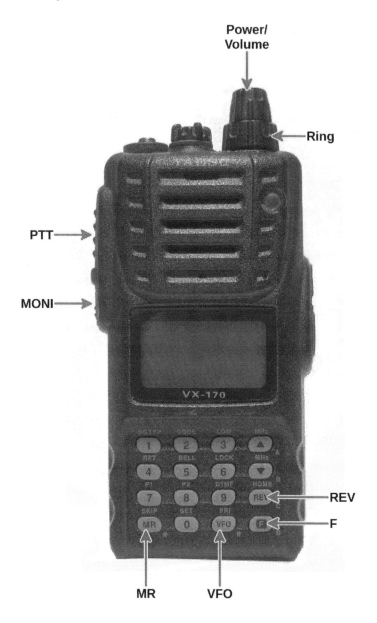

Power/Volume

Ring

PTT

MONI

REV

F

MR

VFO

Specs

Receivers Single receiver with priority channel option
Receives 137–174 MHz FM
Transmits 144–148 MHz @ 5 W FM
Antenna connector SMA F on radio; needs SMA M antenna
Modes FM
Memory Channels 200
Power 6.0–16.0 V DC, EIAJ-02 barrel style, 4mm OD, 1.7mm ID plug, center positive
Model year 2005

Standard Tasks

Program frequency in the field

1. If you aren't already in VFO mode, press [VFO]. The display will show -A- or -b- in VFO mode, and a memory number in memory mode.

2. Set frequency: use the keypad (144390 for 144.390 MHz).

3. Enter menu mode: press [F] [0].

4. Set transmit tone type: use the outer ring to select menu 44:SQL.TYP. Press [F] to adjust. There are six settings: TONE to send tone on transmit, TSQL to send tone on transmit and require tone on receive, REV TN to mute when tone received, DCS digital coded squelch, ECS "enhanced paging," and OFF. Use the outer ring to select, and press [F] when done.

5. Set transmit tone: use the outer ring to select menu 46:TN FRQ for CTCSS or menu 13:DCS.COD for DCS. Press [F] to adjust value. Use the outer ring to select, and press [F] when done.

6. Set repeater shift: use the outer ring to select menu 35:RPT.MOD. Press [F] to adjust. Use the outer ring to select from RPT. +, RPT. -, and RPT.OFF (simplex). Press [F] when done.

7. If you need to change the repeater offset frequency: use the outer ring to select menu 41:SHIFT. Press [F] to adjust. Use the outer ring to select the value. Press [F] when done.

8. Exit menu mode: press [PTT] (does not transmit).

9. Set transmit power: press [F] [3]. Use the outer ring to select the level you want: HIGH (5 W), MID (2 W), or LOW (0.5 W). Press and release [F] to set.

10. Write to a memory: hold [F] for at least one second.

11. Use the outer ring to select the desired memory within five seconds.

12. Press and release [F] to write.

13. Go to memory mode: press [MR].

14. Select the memory you just wrote: use the outer ring.

Lock/unlock radio

Press [F] [6] to lock/unlock.

Check repeater input frequency

Press and release [REV] to switch between reverse and regular modes.

Change power in the field

To set transmit power, press [F] [3]. Use the outer ring to select the level you want: HIGH (5 W), MID (2 W), or LOW (0.5 W). Press and release [F] to set.

Adjust volume

Rotate the center power/volume knob to adjust volume.

Adjust squelch

Press [F] [MONI]. Rotate the outer ring to adjust squelch. Press [F] to set.

Weird Modes

Radio shows OUT.RNG

This means you have ARTS mode enabled. Press [F] to exit ARTS mode.

Can't enter VFO mode

This radio has a memory-only mode. To enable/disable it, hold [MONI] while turning the radio on. Use the outer ring to select F5 M-ONLY and press [F].

Can't change repeater shift

This radio has an automatic repeater shift setting. To enable/disable it, press [F] [0]. Use the outer ring to select 4:ARS. Press [F] to change, and use the outer ring to select ARS.OFF or ARS. ON. Press [F] to set.

Useful Information

[F] [0] menu 12:DC VLT [F] will show battery info.

Factory reset

Hold [MONI] while turning the radio on. Use the outer ring to select F4 ALLRST (reset everything). Press [F] to reset.

Memory reset

Hold **MONI** while turning the radio on. Use the outer ring to select `F2 MEMRST` (clear memories). Press **F** to reset.

Memory bank reset

Hold **MONI** while turning the radio on. Use the outer ring to select `F3 MB RST` (clear memory banks). Press **F** to reset.

Settings reset

Hold **MONI** while turning the radio on. Use the outer ring to select `F1 SETRST` (reset settings). Press **F** to reset.

Yaesu VX-1R

Radio Layout

Knob

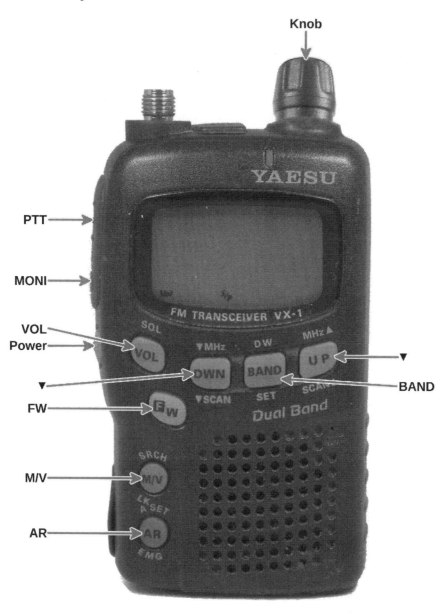

PTT

MONI

VOL
Power

FW

M/V

AR

BAND

Specs

Receivers Single receiver, dual watch (first to break squelch wins)
Receives 0.5–1.7 MHz AM, 76–108 MHz FM, 108–137 MHz AM, 137–999 MHz FM less cellular
Transmits 144–148 MHz @ 0.5 W FM, 430–450 MHz FM @ 0.5 W
Antenna connector SMA F on radio; needs SMA M antenna
Modes FM
Memory Channels 52 in group 1, 142 in group 2
Power 3.2–7.0 V DC, EIAJ-01 barrel style, 2.35mm OD, 0.70mm ID plug, center positive
Model year 1997

Standard Tasks

Program frequency in the field

1. If you aren't already in VFO mode, press **M/V**. The screen will show a channel above the frequency in memory mode, but not in VFO mode.

2. If needed, change band: press **BAND** to select either V-HAM (2 m) or U-HAM (70 cm).

3. Set frequency: use the knob. Press **FW** to change in 1 MHz steps; press **FW** again to go back to normal.

4. Enter menu mode: hold **BAND**.

5. Set transmit tone: press **▲** or **▼** to select menu 25:T SET or menu 27:DCS. Use knob to adjust value.

6. Set transmit tone type: press **▼** to select menu 24:SQL TYP.. Use the knob to select T (tone), T SQ (tone squelch) or DCS.

7. If you need to change the repeater offset frequency: press **▼** to select menu 6:SHIFT. Use the knob to adjust. Note that this affects VFO shift as well.

8. Set repeater shift: press **▲** **▼** to select menu 5: RPTR. Use the knob to select + (positive), - (negative) or blank (simplex).

9. Set transmit power: press **▲** **▼** to select menu 1:TX PWR. Use the knob to adjust the value to HIGH (0.5 W) or LOW (50 mW).

10. Exit menu mode: press **BAND**.

11. Hold **FW** to enter memory write mode.

12. Use the knob to select the memory to write into. Press **FW** to write.

13. Press **M/V** to get to memory mode; scroll to the memory you just wrote.

Lock/unlock radio

Hold [M/V] for one second to lock/unlock.

Check repeater input frequency

Press [FW] and hold [MONI]. Releasing [MONI] switches back to normal mode.

Change power in the field

To set transmit power, Hold [BAND] for one second to enter menu mode, then use [▲][▼] to select menu 1:TX PWR. Use the knob to adjust the value to HIGH (0.5 W) or LOW (50 mW). Press [BAND] to set.

Adjust volume

Press [VOL], then within two seconds use knob to set volume (1–31 or MUTE). Volume mode automatically exits after two seconds.

Adjust squelch

The radio is in "Auto-squelch" mode by default. To adjust, press [FW] then [VOL]. Then within two seconds use the knob to adjust from SQL AUT (auto), SQL OPEN (open) and squelch values 1–10. Squelch mode automatically exits after two seconds.

Weird Modes

Knob doesn't change memory

This radio's knob can either adjust volume/squelch or change memory. By default it selects memory, but you can use 29:DIAL M to switch it to VOL/SQL or DIAL.

Radio is beeping

The radio has an Emergency Mode that you enter by holding [AR] for two seconds. Turn the radio off then on again to exit.

Can't change repeater shift

This radio has automatic repeater shift. To enable/disable this feature, hold [BAND] to enter the menu. Press [▲][▼] to select menu 4: ARS and use the knob to select ARS ON or ARS OFF. Press [BAND] to set.

Radio shows OUTRNG or IN RNG

This means you have ARTS mode enabled. Press [AR] to exit ARTS mode.

Useful Information

Hold [Power] for half a second to turn the radio on or off.

Menu 14: LOCK controls what the radio can do when locked. Options are KEY key lock, DIAL knob lock, D + K (knob and key lock), PTT (PTT lock—no transmit), K + P (key + PTT lock), D + P (knob + PTT lock) and ALL (everything locked).

Memory group 2 is intended for simplex frequencies. It can store repeater shift and tone mode but not tone frequency.

The [MONI] button looks like it's part of the [PTT] button. It's really a separate button.

Factory reset

To reset the radio entirely (including memories), hold [M/V] and [AR] while turning the radio on, then release all buttons. Display will show INI? F. Press [FW] to reset. Display will show INITIAL and then reset.

Settings reset

To reset the radio to default settings (does not affect memories), hold [FW] and [VOL] while turning the radio on, then release all buttons. Display will show SETINN and then reset.

Yaesu VX-6R

Radio Layout

Knob

Ring

PTT

Power

F/W

HM/RV

⊠

MODE

V/M

Specs

Receivers Single receiver
Receives 0.5–998.900 MHz
Transmits 144–148 MHz @ 5 W FM, 222–225 MHz @ 1.5 W FM, 430–450 MHz @ 5 W FM
Antenna connector SMA F on radio; needs SMA M antenna
Modes FM
Memory Channels 1000
Power 5–16 V DC, EIAJ-02 barrel style, 4mm OD, 1.7mm ID plug, center positive
Model year 2005

Standard Tasks

Program frequency in the field

1. If you aren't already in VFO mode, press **V/M**. The screen will show a channel number in memory mode, no channel in VFO mode.

2. Set frequency: use the keypad (144390 for 144.390 MHz).

3. Enter menu mode: press **F/W** then **0** (SET).

4. Set transmit tone type: rotate ring to `60: SQL TYPE`. Press **0** (SET) then use ring to select from `OFF` (no tones), `TONE` (transmit tone), `TSQL` (transmit and receive tones) and `DCS`. Press **0** (SET) to set.

5. Set transmit tone: rotate ring to `66: TN FREQ`. Press **0** (SET) to adjust, then rotate ring to select tone. Press **0** (SET) to set.

6. Set repeater shift: rotate ring to select `51: RPT`. Press **0** (SET) to adjust, then rotate ring to select `-RPT` (negative), `+RPT` (positive) and `SIMP` (simplex). Press **0** (SET) to set.

7. If you need to change the repeater offset frequency: rotate ring to `56: SHIFT`. Press **0** (SET) to change, then rotate ring to select value (0.60 is 600 kHz). Press **0** (SET) to set.

8. Exit menu mode: press **PTT** (does not transmit).

9. Set transmit power: press **F/W** then **⊠** to begin setting power level. Then press **⊠** to select from `LOW1` (0.3 W, 0.2 W on 220), `LOW2` (1 W, 0.5 W on 220), `LOW3` (2.5 W, 1.0 W on 220) and `HIGH` (5 W, 1.5 W on 220). Wait three seconds to leave power setting mode.

10. Write to a memory: hold **F/W** for at least one second. `F` will flash.

11. Rotate ring to select desired memory (flashing channel number means empty). Press and release **F/W** to write.

12. Go to memory mode: press **V/M**.

13. Select the memory you just wrote: rotate the ring.

Lock/unlock radio

Press and hold ⌗ for at least two seconds to lock/unlock. While unlocking, the radio will play a tone that you can ignore. Note that the lock style can be adjusted.

Check repeater input frequency

Press and release HM/RV to switch between reverse and regular modes. Shift direction will flash when in reverse mode.

Change power in the field

To set transmit power, press F/W then ⌗ to begin setting power level. Then press ⌗ to select from LOW1 (0.3 W, 0.2 W on 220), LOW2 (1 W, 0.5 W on 220), LOW3 (2.5 W, 1.0 W on 220) and HIGH (5 W, 1.5 W on 220). Wait three seconds to leave power setting mode.

Adjust volume

Rotate the knob to adjust volume.

Adjust squelch

Press F/W then 0 (SET) to enter set mode. Use ring to select 59: SQL. Press 0 (SET) to change. Rotate ring to choose from LVL 0 (no squelch) to LVL 15. Press 0 (SET) to set, then PTT to leave menu mode.

Weird Modes

If someone says you have WIRES mode turned on

You are transmitting a DTMF tone that interrupts your first few syllables. Press and release the ⌗ key once. The ⌗ symbol on the right of the display will disappear.

Radio shows OUT RANGE

This means you have ARTS mode enabled. Press 4 (ARTS) to disable.

Useful Information

Hold Power for one second to turn the radio on or off.

Be careful when adjusting the transmit power or locking/unlocking that you do not inadvertently turn WIRES mode on. Make sure the ⌗ symbol is not visible in the lower left.

Menu 35: LOCK lets you select from KEY, DIAL, K+D, PTT, P+K, P+D, and ALL for the lock mode.

Factory reset

To reset everything, hold [0] (SET), [MODE] and [V/M] while turning the radio on, then press [F/W] to reset.

Settings reset

To reset all menu settings but leave memories intact, hold [MODE] and [V/M] while turning the radio on, then press [F/W] to reset.

Radio Layout

Specs

Receivers Two independent receivers, simultaneous receive

Receives 0.5–30 MHz, 50–999 MHz on main band, 50–54 MHz, 140–174 MHz, 420–470 MHz on sub band

Transmits 50–54 MHz @ 5 W FM, 144–148 MHz @ 5 W FM, 420–470 MHz @ 5 W FM, 222–225 MHz @ 0.3 W FM, 50–54 MHz @ 1 W AM

Antenna connector SMA F on radio; needs SMA M antenna

Modes FM, AM (main VFO only)

Memory Channels 450

Power 10–16 V DC, EIAJ-02 barrel style, 4mm OD, 1.7mm ID plug, center positive

Model year 2002

Standard Tasks

Program frequency in the field

1. If you aren't already in VFO mode, press [V/M]. The screen will show VFO when you are in VFO mode.

2. Set frequency: use the keypad (144390 for 144.390 MHz).

3. Enter menu mode: press [MON F] then [0](SET).

4. Set transmit tone type: rotate ring to select TSQ/DCS/DTMF:1 SQL TYPE. Press [MAIN]/[SUB] to cycle through OFF, DCS, TONE SQL (transmit and receive tones) and TONE (transmit tone only).

5. Set transmit tone: rotate ring to select TSQ/DCS/DTMF:2 TONE SET. Press [BAND] to adjust, then use [MAIN]/[SUB] to select tone. Press [BAND] when done. To select DCS tone, use TSQ/DCS/DTMF:3 DCS instead.

6. Set repeater shift: rotate ring to select Basic Setup:7 RPT SHIFT. Press [MAIN]/[SUB] to cycle through +RPT (positive shift), -RPT (negative shift) and SIMP (simplex).

7. Exit menu mode: press [MON F][0](SET).

8. Set transmit power: press [MON F] then [⌧]. Repeatedly press [⌧] to select from L1 (0.05 W), L2 (1 W), L3 (2.5 W) and no indication (5 W). Press and release [MON F] to save.

9. Write to a memory: hold [MON F] for at least half a second. Use ring to select the desired memory (* means empty). Press and release [MON F] to write.

10. Go to memory mode: press [V/M].

11. If you need to change the repeater offset frequency: see the "Useful Info" section.

Lock/unlock radio

Press and hold [⊗] for at least two seconds. Note that you can adjust the lock mode with menu `Basic Setup:10 LOCK MODE`.

Check repeater input frequency

Press and release [HM/RV] to switch between reverse and regular modes.

Change power in the field

To set transmit power, press [MON F] then [⊗]. Repeatedly press [⊗] to select from L1 (0.05 W), L2 (1 W), L3 (2.5 W) and no indication (5 W). Press and release [MON F] to save.

Adjust volume

Rotate the knob to adjust volume.

Adjust squelch

Press [MON F] [0] (SET) to enter the menu. Use the ring to select `Basic Setup:1 SQL NFM`. (The radio also has `Basic Setup:2 SQL WFM`.) Press [MAIN]/[SUB] to adjust. Press [MON F] [0] (SET) to leave the menu. Squelch is separate for main and sub bands.

Weird Modes

If someone says you have WIRES mode turned on

You are transmitting a DTMF tone that interrupts your first few syllables. Press and release the [⊗] key once. Radio/PC/Globe symbol will disappear.

Radio shows OUT RANGE

This means you have ARTS mode enabled. Press [MON F] then [4] (ARTS) to disable.

Radio is beeping and LED is flashing

You have entered emergency mode, probably by holding [HM/RV] for two seconds. Turn the radio off and on again to disable.

Useful Information

Hold [Power] for two seconds to turn the radio on or off.

The order of menu sections is `Basic Setup` (1–14), `Display Setup` (1–8), `TSQ/DCS/DTMF` (1–8), `Scan Modes` (1–7), `Measurement` (1–7), `Save Modes` (1–6), `ARTS` (1–3), `Misc Setup` (1–20).

Be careful when adjusting the transmit power or locking/unlocking the radio that you do not inadvertently turn WIRES mode on. Make sure the Radio/PC/Globe symbol is not visible in the upper-left.

You must disable automatic repeater shift in order to set repeater shift and offset. To do this, press **MON F** **0** (SET) to enter menu mode. Then use the ring to select `Basic Setup:5 ARS` and use **MAIN**/**SUB** to adjust. Press **MON F** **0** (SET) to leave menu mode.

There are two ways to change repeater offset. The first, which changes the repeater offset globally for all memories in the same band, is to press **MON F** **0** (SET) to enter menu mode. Then use the ring to select `Basic Setup:6 SHIFT` and use **MAIN**/**SUB** to adjust. Press **MON F** **0** (SET) to leave menu mode.

If you need to change an offset for only one memory (odd split), instead program the memory as normal with the receive frequency. Then go back to VFO mode, tune to the transmit frequency, and hold **MON F** for half a second. Select the memory with the odd split using the ring, and hold **PTT** (does not transmit). Press **MON F** to save the transmit frequency into the same memory.

This radio comes in two colors: black and silver.

Factory reset

To reset everything, hold **4**, **BAND** and **V/M** while turning the radio on.

Settings reset

To reset all settings but leave memories intact, hold **BAND** and **V/M** while turning the radio on.

Yaesu VX-8DR

Radio Layout

Knob

PTT

MONI

VOL

F/W

MENU

BAND

HM/RV

⌀

Power

V/M

Specs

Receivers Two independent receivers, simultaneous receive
Receives 510–1790 kHz AM on VFO A, 0.5–30 MHz AM on VFO A, 30–76MHz FM
on VFO A, 108–999.9 MHz FM on VFO A (cell blocked), 108–580 MHz FM on
VFO B
Transmits 50–54 MHz @ 5 W FM, 144–148 MHz @ 5 W FM, 440–450 MHz @5
W FM, 222–225 MHz @1.5 W FM, 50–54 MHz @ 1 W AM
Antenna connector SMA F on radio; needs SMA M antenna
Modes FM, AM (VFO A only)
Memory Channels 900
Power 11–14 V DC, EIAJ-02 barrel style, 4mm OD, 1.7mm ID plug, center positive
Model year 2009

Standard Tasks

Program frequency in the field

1. If you aren't already in VFO mode, press [V/M]. The screen will show VFO in
   the upper left in VFO mode.

2. Set frequency: use the keypad (144390 for 144.390 MHz).

3. Enter menu mode: hold [MENU] for at least half a second.

4. Set transmit tone: use the knob to select 99:TONE FREQUENCY then press
   [MENU] to adjust. Use the knob to adjust, and press [MENU] to set.

5. Set transmit tone type: use the knob to select 95:SQL TYPE then press [MENU]
   to adjust. Use the knob to select from TONE (transmit tone), TONE SQL (transmit
   and receive tone), DCS (digital squelch), REV TONE (mutes when tone received),
   PR FREQ (mutes when PR frequency tone is received), PAGER (pager tone
   received), MESSAGE (APRS message recieved) or OFF. Press [MENU] to set.

6. Set repeater shift: use the knob to select 75:RPT SHIFT. Press [MENU] to
   adjust. Use the knob to select from -RPT (negative), +RPT (positive) or SIMPLEX
   (simplex), then press [MENU] to set.

7. Exit menu mode: hold [MENU] for at least half a second.

8. Set transmit power: press and release [F/W]. Press and release [⊠] (TX PO)
   repeatedly to select power level. Options are L1 (50 mW), L2 (1.5 W), L3 (2.5
   W) and HI (5 W). Press and release [F/W] to set.

9. Write to a memory: hold [F/W] for at least half a second.

10. Use the knob to select desired memory.

11. Press and release [F/W] to write. If prompted with OVERWRITE? press [F/W] again
    to confirm.

12. Go to memory mode: press [V/M].

13. Select the memory you just wrote: use the knob.

14. If you need to change the repeater offset frequency: see the "Useful Info" section.

Lock/unlock radio

Press and release [Power] button for less than half a second to lock/unlock.

Check repeater input frequency

Press and release [HM/RV] to switch between reverse and regular modes.

Change power in the field

To set transmit power, press and release [F/W]. Press and release [⊗] (TX PO) key repeatedly to select power level. Options are L1 (50 mW), L2 (1.5 W), L3 (2.5 W) and HI (5 W). Press and release [F/W] to set.

Adjust volume

Hold [VOL] and rotate the knob to adjust. Each VFO has its own volume.

Adjust squelch

Press [F/W] [MONI]. Use the the knob to change level (0–15), and press [MONI] to set.

Weird Modes

If someone says you have WIRES mode turned on

You are transmitting a DTMF tone that interrupts your first few syllables. Press and release the [⊗] key once. Atom symbol will disappear.

Radio shows OUTRNG

This means you have ARTS mode enabled. Press and release [4] (ARTS) to disable.

Bar graph shows across the screen

You have entered the spectrum analyzer mode. Press [V/M] then [F/W] [8] (SP ANA) to exit.

Can't leave mode

If you enter a mode that let you set memory values by using a keypad shortcut, you need to press the same button to leave the mode. For example, if you press [F/W] [6] (RPT) to set repeater shift, you need to push [6] to leave the mode.

Radio volume control stuck or reversed

The VX-8 has a feature for mobile use that will reverse how the volume adjustment behaves. If this feature is enabled, the knob and arrows will adjust volume but not frequency. Holding down the volume control with this feature enabled will provide "normal" behavior. To get in or out of this mode, press and release the [F/W] key, then press and release the [VOL] key.

Radio shows APRS information

The VX-8 will enter APRS mode if you press (not hold) [MENU]. The first press shows your position. Pressing again displays a STATION LIST screen. Pressing it a third time will show APRS MESSAGE. Pressing it a fourth time will take you back to the normal mode.

Menu doesn't have correct options

This radio has two menu systems, one for regular system settings and one for APRS. If you hold [MENU] while in APRS mode, you'll see an APRS menu. Hold [MENU] to exit the APRS menu, then press [MENU] multiple times as described in the previous topic to exit APRS mode. Finally, hold [MENU] to get back to the normal menu.

Useful Information

Hold [Power] for half a second to turn the radio on or off.

Be careful when adjusting the transmit power that you do not inadvertently turn WIRES mode on. Make sure the ⊗ symbol is not visible in the lower left.

The radio can have GPS installed on radio body (mic connector) or on speaker mic. If you don't need it, you can remove it to save battery (needs a screwdriver).

Hold the button for VFO A or VFO B to switch between monitoring two frequencies and monitoring one frequency.

There are two ways to change repeater offset. The first, which changes the repeater offset globally for all memories in the same band, is to hold [MENU] for at least half a second to enter menu mode. Then select 76:RPT SHIFT FREQ and press [MENU]. Use the knob to change the frequency and press [MENU] to set.

If you need to change an offset for only one memory (odd split), instead program the memory as normal with the receive frequency. Then go back to VFO mode, tune to the transmit frequency, and hold [F/W] for half a second. Select the memory with the odd split using the knob, and hold [PTT] (does not transmit). Press [F/W] to save the transmit frequency into the same memory.

Factory reset

To reset everything, hold [BAND], [HM/RV] and [⊗] while turning the radio on, then press [F/W].

Settings reset

To reset most settings but leave memories intact, hold **BAND** and **V/M** while turning the radio on, then press **F/W**.

Radio Layout

Power/
Volume

PTT

VFO/MR

MENU

EXIT

#

Specs

Receivers Single receiver, dual watch (first to break squelch wins)
Receives 136–174 MHz and 400–480 MHz FM
Transmits 136–174 MHz @ 4 W FM and 400–480 MHz @ 4 W FM
Antenna connector SMA **M** on radio; needs SMA **F** antenna
Modes FM
Memory Channels 128
Power No DC input on radio
Model year 2012

Standard Tasks

Program frequency in the field

1. This radio doesn't allow you to overwrite memories. You will need to delete first if you want to write to a memory that has data in it. To delete: press MENU 2 8 MENU (DEL-CH) *XXX* where *XXX* is the channel (001–128). Then press MENU.

2. If you aren't already in VFO mode, press VFO/MR to go to VFO mode if you aren't there already. The screen will show a channel number if you are not in VFO mode.

3. Set frequency: use the keypad (144390 for 144.390 MHz). You may need to press BAND to switch bands depending on your frequency.

4. Set transmit tone type and value: press MENU 1 3 MENU (T-CTCS) then use ▲ ▼ to select correct CTCSS tone frequency (or OFF). Press MENU EXIT.

5. Set repeater shift: press MENU 2 5 MENU (SFT-D) and use ▲ ▼ to select the correct repeater shift. Press MENU EXIT.

6. If you need to change the repeater offset frequency: press MENU 2 6 MENU (OFFSET) and enter the value with the keypad (00600 for 600 kHz, 05000 for 5 MHz). Press MENU EXIT.

7. Set transmit power: press MENU 2 MENU (TXP) and use ▲ ▼ to select power level: LOW is 1 W, HIGH is 4 W. Press MENU EXIT.

8. Write to a memory: press MENU 2 7 MENU (MEM-CH. Enter channel to write *XXX* (001–128) then press MENU EXIT.

9. Go to memory mode: press VFO/MR.

10. Select the memory you just wrote: use ▲ ▼.

Lock/unlock radio

Hold # for three seconds to lock/unlock.

Check repeater input frequency

This radio has no option to listen to the repeater input frequency. You will have to program a separate memory with the repeater input frequency to do this.

Change power in the field

To set transmit power, press [MENU] [2] [MENU] (TXP) and use [▲]/[▼] to select power level: LOW is 1 W, HIGH is 4 W. Press [MENU] [EXIT].

Adjust volume

Rotate power/volume knob to adjust volume.

Adjust squelch

Press [MENU] [0] [MENU] (SQL) then scroll to the correct squelch level, press [MENU] [EXIT].

Weird Modes

Can't leave channel (memory) mode

Some distributors ship these radios with VFO mode turned off. If that's the case, you will need to modify the programming with a computer/radio programming cable to turn VFO mode on. This cannot be changed from the front panel.

Can't set offset/direction

This radio can display frequencies instead of channel names. If you have that enabled, it's difficult to tell if you're in channel (memory) mode or frequency (VFO) mode. One sure way is to try to program an offset or shift. If it doesn't "stick" after programming, that's likely because you're in channel mode. Switch to frequency mode in order to program.

Useful Information

This appears to be a re-badged Baofeng UV-5R with modified firmware.

Hold [3] while turning the radio on to see the firmware version.

If you wait too long after pressing [MENU] it will time out. You need to press quickly.

No reset procedure

This radio cannot be reset from the front panel.

⚠ WARNING: In at least some configurations, the Zastone ZT-V8 may permit you to transmit on business or public safety frequencies. Make sure you are in-band when transmitting.

Acknowledgments

It's very easy to end up owning too many handhelds. Thanks to the following people, who very graciously allowed me to use their handhelds and let me mess up their programming in the process of writing this book:

Michael Drapkin WB2SEF lent me his Icom IC-P2AT and Icom IC-2AT. (Or maybe one of them was his wife Jennifer N7ZID's radio that Michael lent me.)

Tsuyoshi Nagano WH7JJ lent me his Kenwood TH-D72A.

Rich Langevin KB7YEB lent me his Kenwood TH-D7 and Yaesu FT-50.

Ed Karsten K9EDK gave me his Icom IC-W32A.

Jim Pierce N7QVW lent me his Icom IC-T70A and Yaesu VX-170.

Dennis Bietry KE7EJF let me play with his Harris Unity XG-100P.

Julie Jennings KI6UCJ lent me her Kenwood TH-F6a.

Walt Reinert NJ8G lent me his Yaesu VX-6R.

Phil Arnold KJ6STS lent me his Kenwood TH-78A.

Arnold Knack KA7AOK let me photograph his Icom ID-51A.

Ray Novak N9JA and Icom America lent me an Icom ID-31A PLUS.

Robert Fairfield K7RQN lent me his Kenwood TH-215A and TH-315A.

Mitch Sprowl K1MAS let me play with his Yaesu FT-1DR.

Thanks to the great people at Ham Radio Outlet in Phoenix, who let me photograph and play with the the Yaesu FT-70D, Yaesu FT-2D, Icom IC-V80 HD, Kenwood TH-K20A, Alinco DJ-500T, Kenwood TH-D74, Yaesu FT-25, Yaesu FT-65, Yaesu FT-270, Tera TR-590, Wouxun KG-UV9D, Hytera AR482G, Yaesu FT-3D, and Icom IC-V86. In particular I'd like to thank manager Ron McKee AJ7T, former manager Mike Spraker KB6VHF, Jim Skinnell K7EOG and Mike Baker K7DD. If you're lucky enough to have a local ham store, please support it!

Thanks to everone whose suggestions helped improve the book. Jim Pierce N7QVW suggested the chapters on batteries and on radios before they were programmable. Andy Durbin K3WYC gave me the tip on Wouxun dealer mode. Michael Drapkin WB2SEF suggested the glossary chapter and provided a careful review. Rob Rowlands NZ6J pointed out some issues in the Baofeng UV-5R, UV-82, Yaesu FT-2D and FT-3D chapters.

Thanks also to everyone in the Maricopa County Emergency Communications Group, for providing a way for hams to learn about their radios in the field while helping others.

Thanks to Renée Targos for her help with editing. Any spelling errors you don't see are due to her.

Thanks to the creators of TEXstudio, LaTeX, the Memoir style, TikZ, TEX and GIMP for making the wonderful tools I used for this book.

Thanks to my father, Andy, for leaving the license manuals at my house and leading me into ham radio. Thanks to my mother, Nancy, for giving me the love of books that I've always had—and without which this book would never have existed.

And finally, thanks to my wife, Debbie, who still likes it when I put antennas in the yard.

Let's keep in touch!

This book comes from my experience, mostly in public service events. The first spark was lit when I was participating in my second public service event with a brand new Yaesu VX-8R. Somehow I got my radio into a weird mode (which I now know to be ARTS) and didn't know how to get it out. Since my radio was relatively new, there was no local expert who could guide me. Turning the radio off and on again was no help. As the minutes counted down to the race start, I began hitting all the keys on my keypad systematically. Finally I got myself out of that cursed mode.

The spark burst into a flame at a later event. I had programmed the portable repeater's frequency using programming software the night before. When the portable repeater ended up on a different frequency, I wasn't able to get my radio reconfigured properly. Embarrassingly, I had to borrow another ham's backup radio to participate in the event. After that, I created quick and dirty programming instructions for all my radios. I kept them with my go bag.

Those public service events taught me a lot about my radio. I realized that *every* ham faced similar issues with program radios on the fly and recovering from configuration problems. My notes grew, and this book is the result.

I'd like to make this book the best it can be. To do that, I need your help! Are there things that you think could be described better? Did you run into any errors or unclear instructions? Are there weird modes you know of that aren't described in the book?

If you think something was good or not so good, I want to hear from you. If some part of this book has helped you out, I'd love to hear your story!

You can contact me at andrew@handheldradio.net.

If you *own* a radio that's not included and are willing to part with it for a brief period, please get in touch! I'm always looking for new radios and would love to add your name to the Acknowledgments.

You can sign up at handheldradio.net for the Handheld Radio Field Guide email list. I'll send out updated chapters when I fix errors and let you know when new editions become available.

If you think this book is valuable, a five-star review on Amazon helps bring it to the attention of other hams. As soon as Amazon or Ingram offers a lay-flat spiral binding option, I will too!

Not every ham thinks it's important to be able to program your radio from the front panel. I'm glad you do.

References

[1] Coaxial Power Connector. In *Wikipedia*. Retrieved from https://en.wikipedia.org/wiki/Coaxial_power_connector

[2] *ECR-5892D: Comparison of NiCd, NiMH, and Li-Ion Batteries*. Retrieved from https://www.portal.state.pa.us/portal/server.pt/directory/technical_publications

[3] EIAJ Connector. In *Wikipedia*. Retrieved from https://en.wikipedia.org/wiki/EIAJ_connector

[4] Emission Classification. In *Amateur Radio Wiki—The Online Encyclopedia for Hams*. Retrieved from http://www.amateur-radio-wiki.net/index.php?title=Emission_Classification

[5] *FCC ID Application Database*. http://fccid.io

[6] *Hampedia*. http://hampedia.net

[7] *Internet Archive*. http://archive.org

[8] *RigPix*. http://www.rigpix.com

[9] *RigReference.com*. http://rigreference.com

[10] *SARL Radio Age and Price Guide*. http://www.sarl.org.za/Radios.asp

[11] Understanding the Radioshack Adaptaplug System. In *RadioShack Guide to Understanding Power Conversion*. Retrieved from http://support.radioshack.com/support_tutorials/batteries/pwrgde-2H.htm

Made in the USA
Las Vegas, NV
13 February 2021

17796437R00188